THE LANGUAGE OF THIS LAND,

MI'KMA'KI

*To Bob — another
author who is
keen on the
written word!
Berni*

Cape Breton University Press
Sydney, Nova Scotia

THE LANGUAGE OF THIS LAND,

MI'KMA'KI

TRUDY SABLE AND BERNIE FRANCIS

WITH
WILLIAM JONES
ROGER LEWIS

FOREWORD BY LEROY LITTLE BEAR
CHAIR, AMERICAN NATIVE STUDIES
UNIVERSITY OF LETHBRIDGE

Cape Breton University Press
Sydney, Nova Scotia

This book is a product of a lineage of Elders who spoke this language, danced these dances and walked this land, Mi'kma'ki. Through their generosity and courage, despite decades of cultural disruption, discrimination and marginalization, they have carried their traditions forward with humour and generosity. This is the greatest gift left for all people living in Mi'kma'ki today.

Cape Breton University Press recognizes the support of Canada Council for the Arts and of the Province of Nova Scotia, through the Department of Communities, Culture and Heritage. We are pleased to work in partnership with these bodies to develop and promote our cultural resources.

Canada Council   Conseil des Arts
for the Arts   du Canada

NOVA SCOTIA
Communities, Culture and Heritage

Cover design: Cathy MacLean, Pleasant Bay, NS
Cover image: Cape Split. Photo by David Sable

Layout: Mike Hunter, Port Hawkesbury and Sydney, NS
First printed in Canada

Library and Archives Canada Cataloguing in Publication

Sable, Trudy, 1952-
The language of this land, Mi'kma'ki / Trudy Sable
and Bernie Francis.

Includes bibliographical references.
ISBN 978-1-897009-49-9

1. Micmac Indians--Nova Scotia--Folklore. 2. Micmac
Indians--Nova Scotia--Religion. 3. Indian dance--Nova Scotia.
4. Folk songs, Indian--Nova Scotia. 5. Legends--Nova Scotia.
6. Geology--Nova Scotia--Folklore. 7. Nova Scotia--Folklore.
8. Nova Scotia--Social life and customs. I. Francis, Bernard,
1948- II. Title.

E99.M6S22 2012 398.2089'97343 C2012-900410-3

Cape Breton University Press
P.O. Box 5300
Sydney, NS B1P 6L2

TABLE OF CONTENTS

FOREWORD

By Leroy Little Bear
University of Lethbridge

Blackfoot think of time on a two-day operational sense. There is "now," "tomorrow," and "day-after tomorrow." And backwards, "now," "yesterday," and "day-before yesterday." Beyond the two-day limit, forward or backward, past and present amalgamate and become one and the same. Plains Indians are not incapable of talking or thinking of the distant future or past, but it is always done with the "constant flux" in mind. One of the implications arising out of this notion of time is that the ancestors are always only two days away. The stories, the songs, the ceremonies, the teachings are never more than two days old in the memory of the people. This is quite different from Pierre Elliott Trudeau's statement, typifying Euro-Canadian worldview, to the effect of "these treaties are not worth the paper they are written on." In other words, what is past is past … it is gone forever. The only thing that matters is the future. (Little Bear 2001: 5)

When I was approached to make a contribution to *The Language of this Land, Mi'kma'ki* in terms of writing a foreword, I accepted without hesitation. I accepted without hesitation for two reasons: the book explains the world view and language of the Mi'kmaq, similar to a life-long endeavour of this writer regarding the Blackfoot. Secondly, the book strikes at the heart of the "humanities" in academia. At the risk of repeating myself from other writings and presentations, I would like to share some thoughts with the reader as I did with participants at a symposium on the humanities.

This author has always understood humanities as being about humans and their culture including philosophy, worldview and arts and excluding hard sciences. A definition of the humanities that strikes a deep chord is one framed by the Ohio Humanities Council as follows:

The humanities are the stories, the ideas, and the words that help us make sense of our lives and our world. The humanities introduce us to people we have never met, places we have never visited, and ideas that may have never crossed our minds. By showing how others have lived and thought about life, the humanities help us decide what is important in our own lives and what we can do to make them better. By connecting us with other people, they point the way to answers about is right or wrong, or what is true to our heritage and our history. The humanities help us address the challenges we face together in our families, our communities, and as a nation.

The above definition of the humanities captures very well the experience of Aboriginal peoples in academia. In other words, Aboriginal peoples are forever explaining themselves to non-Aboriginal people: telling their stories, explaining their beliefs and ceremonies, and introducing ideas that, in many cases, have never crossed the non-Aboriginal mind. The following overview of traditional knowledge serves as a good example of what is meant.

Edward H. Spicer refers to Daniel Corkery to make a point about North American Indians. Corkery, in 1925, had written a book called *The Hidden Ireland*. In it he points out that there existed a rich culture in Ireland hidden from the English colonizers.

> Few, if any, of the English people were aware of the Irish language, and those who were aware during the eighteen and nineteenth centuries associated it only with illiterate and backward peasantry. [...] That it had continued into the nineteenth century as a medium of literary life in Ireland was simply beyond the pale of English consciousness." (Spicer 1980)

In other words, Irish culture was hidden from the English. Spicer points out, for instance, that:

> In the United States, the Iroquois and the 'Americans' had known each other for more than two hundred years, but the Americans who made policy for New York State in the 1960s knew nothing of the language, the cultural history or the religion of the Seneca. (Spicer 1980)

Writing in 1973 in *God is Red*, Deloria states:

> Sincere but unknowing whites honestly asked us less than a decade ago if we still lived in tents, if we were allowed to leave the reservations, and other relevant questions, indicating that for a substantial number of Americans, Indians were still shooting at the Union Pacific on their days off. (Deloria 1973: 41)

One can, in a similar manner apply the concept of "hiddenness" to traditional knowledge of Aboriginal peoples in North America. Generally speaking, Euro-Canadians and Euro-Americans know very little of the languages, religions, tradition, sciences and so on, of North American Aboriginal peoples. Traditional knowledge of Aboriginal people has never been taken seriously because it is usually categorized as superstition or folklore. If one were to ask the typical academic about the utility or importance of traditional knowledge in the academic world they would most likely answer, "not much."

There are "tons" of studies of Aboriginal peoples by government, missionaries and social scientists. But the majority if not all of those studies are done from a Eurocentric perspective based on Western paradigms. Only recently has traditional knowledge of Aboriginal people been experiencing a "coming out" in Canada partially brought about by the intellectual property debate and recent Supreme Court of Canada decisions. Ironically, more than twenty years after his earlier observations, Deloria observes:

> When multitudes of young whites roam the West convinced they are Oglala Sioux Pipe Carriers and on a holy mission to protect 'Mother Earth' and when priests and ministers, scientists and drug companies, ecologists and environmentalists are crowding the reservations in search of new rituals, new medicines or new ideas about the land, it would appear as if American Indians finally have it made. (Deloria 1995: 13)

Aboriginal people have for a long time questioned the assumptions and methodologies superimposed on their knowledge systems. Only within the recent past, with the reluctant acceptance of Native studies in academia—and the consequent increase in the number of Aboriginal scholars trained in the Eurocentric tradition—have the writings of Aboriginal scholars challenged Eurocentric assumptions and methodologies and begun to make inroads into academia. This book goes a long way toward filling the gap.

——

The Language of this Land, Mi'kma'ki is a good representation of what the humanities are about. It is the storytelling of what traditional knowledge is about; it is the ideas about reality from a Mi'kmaw perspective; it about ideas that normally do not cross the Western mind; it is an attempt to connect with the Western mind so it will understand the world view, customs and lifeways of peoples like the Mi'kmaq. It resonates with the humanities in academia.

ACKNOWLEDGEMENTS

~Bernie Francis~

Trudy (Tlu') Sable and I have been working together for twenty-one years. During this period, we discussed, argued, fought and worked very hard to attempt to understand the meanings behind many of the legends in the Mi'kmaw world. I had the good fortune of being able to return to the Mi'kmaw Elders to share the discussions Trudy and I had as well as those I had with Ruth Whitehead, and to learn where the Elders stood in regards to these legends. The late Noel Marshall from Potlotek (Chapel Island), Joe B. Marshall, the late Wilfred Prosper and the late Dr. Margaret Johnson, the late Alex Denny and his wife Janet, from Eskissoqnik (Eskasoni), Pauline Bernard, Carl Gould and the late Frank Doucette from Maupeltu (Membertou), Charlie William Francis from Sipekne'katik (Shubenacadie), Charlie Marshall from We'kopekwitk (Millbrook) and the late Caroline Gould from We'kopa'q (Whycocomagh) are some names that come to mind.

The Mi'kmaw language has many layers of meaning discovered through the use of the latest linguistic tools. Along with Trudy's training in anthropology, education and Tibetan/Shambhala Buddhism we were able to do a thorough analysis resulting in this book.

Finally, without the urgings and excellent teaching of Professor Doug Smith who truly introduced to us the science of linguistics, and of Dr. Peter Christmas who had the intestinal fortitude to deal with and act on a very controversial topic (i.e., the need to develop a new orthography reflecting present day Mi'kmaw), perhaps none of this work would have materialized.

~Trudy Sable~

For more than two decades, I have continuously promoted and illustrated how a dialogue could be created within academia, and at all levels of education, between Indigenous Knowledge and our current Western academic discourses. This dialogue would be very powerful and beneficial to understanding our increasingly global and culturally diverse world.

For this reason, it has been a privilege to work with Bernie Francis throughout the last twenty-one years, along with a number of Elders. An earlier version of this work appeared in my MA thesis (1996) entitled, "Another Look in the Mirror: Research into the Foundations for Developing an Alternative Science Curriculum for Mi'kmaw Children." A core chapter in the thesis was on the nature of the Mi'kmaw language, and the world view embedded within it. That research was based on numerous interviews with linguist Bernie Francis and Elders Dr. Margaret Johnson and Wilfred Prosper, along with linguists John Hewson and Doug Smith. Research into the legends, songs, dances and other expressive forms of communicating knowledge, deepened my appreciation for cultural differences in experiencing and expressing reality, all valid in their own right for the purposes of that culture. Now, more than fifteen years later, we are able to expand, correct and delve deeper into this work and offer it in the form of this book.

Sadly, many of the Elders I came to know and/or interview are no longer with us, but their humour, their stories, their kindness and endless cups of powerhouse tea (and subsequent bathroom visits) permeate this book. They hosted or put me up in their houses and put up with me. In particular: I want to acknowledge the major role of Dr. Margaret Johnson, with who I stayed, studied and travelled with for almost twenty years and deeply miss; Wilfred Prosper, whose kitchen became a second home, an incredible learning environment and place of gut-splitting laughter; Caroline Gould, a brilliant, kind and accomplished artist; as well as Noel Doucette, John Basque, Sarah Denny, Rita Joe, Annie Cremo, Checker Bernard, Grand Keptin Alec Denny, Joe Levi Sylliboy, Irene Julian, Annie Battiste, Joe Peters and Frank Doucette to name a few. These Elders never had to proclaim themselves— people just knew who they were and respected them. They have left their indelible legacy for others to follow; they have left the language of this land.

Fortunately, other Elders are still with us. In particular, I want to thank Bessie Prosper, who was part of many of these discussions and who continues to offer a place to stay and her own valuable knowledge. I would also like to express my appreciation to the Grand Chief Ben Sylliboy for the kindness he has shown to me over the years, as well as Georgina Doucette, Doug Knockwood, Greg Johnson, Ellen Robinson, Joe Knockwood, Libby Meuse, Don Julian, Florence Young, Gilbert Sewall, Sandy Julian, John Joe Sark, Isabelle Knockwood, and Noel Knockwood for taking time to speak with me and offer their insights and knowledge at various points in this journey. Others who have supported or helped me in a variety of large and small ways, and sometimes protected me, are Vaughen and Shirley Doucette, Frank Meuse, the families of Dr. Margaret Johnson (Albert, Joan (sadly deceased as well), Elizabeth, Lottie, Mary, Patsy, Carol, Gerald, Gerard, Tom, Prentiss, Derrick, as well as George Paul, and all their spouses, children and grandchildren) and Caroline Gould (Margaret, Marjorie, Annie, Eunice and Ethel), Kenny Prosper, the late Joey Gould, Kerry Prosper, Vivian Basque, Joel Denny, Alan Knockwood, Stephen Ginnish, Ann Ward, Chief George Ginnish, Wallis Nevin, Jim Simon, Lindsay Marshall, Dr. Marie Battiste, Sakej Henderson, Stephen Augustine, Dan Christmas, Eleanor Johnson, Alan Syliboy, the late Basile and Junior Joe, Benjie Lafford, Gary Knockwood, Donna Augustine, the late Tom Paul, Ruby Doucette, Keith Denny, Noel and Lawrence Julian, Eileen Brooks, Dorothy Hache, Beverly Jeddore, Joan Lafford and many others who shared their time, their dances and ceremonies with me along the way.

I am also grateful to Ruth Holmes Whitehead for her continuous friendship and support, and to both her and Harold McGee for their years of dedicated and original research about many aspects of Mi'kmaw culture and history. William Jones, geomatics expert, has contributed enormously to my work for the past decade by generously offering maps illustrating multiple layers of land use and occupancy. Roger Lewis, archaeologist and Assistant Curator of Ethnology at the Nova Scotia Museum, has also contributed his valuable knowledge of the land and river systems, along with his archaeological expertise. Others who have supported or en-hanced my research include: Rob Ferguson, archaeologist (retired) at Parks Canada; David Christianson, archaeologist and Manager, Collections Unit, Museum Operations at the Nova Scotia Museum; Michael Deal, archaeologist, Memorial University of Newfoundland; David Keenlyside, Executive Director of the PEI Museum and Heritage Foundation; Patricia Allen, archaeologist; Ralph Stea, glaciologist; Stephen Davis, archaeologist and Professor Emeritus at SMU; Colin Howell, historian and Professor Emeritus at SMU; Peter MacDonald, former principal at the Eel Ground First Nations School; as well as Jackie Logan, Office Manager of the Gorsebrook Research Institute for her good eye and general support. It goes without saying that my husband David and daughter Julia were of great importance in supporting this book as well as reading and offering many good insights and edits to it themselves, along with the cover photo taken by David.

NOTE FROM THE AUTHORS AND EDITORS

The word Mi'kmaq is plural and is also used when referring to the whole nation. For instance: "The Mi'kmaq of Eastern Canada…."

Mi'kmaw is the singular and adjectival form of Mi'kmaq. Examples: "I am a Mi'kmaw" or "A Mi'kmaw man told me a story" or, "As a Mi'kmaw speaker … etc. "It is also used to refer to the language itself. Examples: "I speak Mi'kmaw. "Mi'kmaw is my first language." "All the Mi'kmaq spoke Mi'kmaw up to the 1950s…."

Mi'kma'ki, the territory of the Mi'kmaq, includes the island of Newfoundland, all of Nova Scotia and Prince Edward Island, much of New Brunswick and the Gaspé, and part of northeastern Maine. Readers should note that while most of the research in this book is concentrated in Nova Scotia, the spirit of the research and knowledge herein applies to the entire region.

Mi'kma'ki: Mi'kmaw territory
Mi'kma'kik: at the Mi'kmaw territory

Many placenames, events, phrases and passages recorded by previous researchers have required reinterpretation in keeping with contemporary orthographic conventions. Where it is important to distinguish, we have noted such improvements (i.e., the Smith Francis Orthography, which has been adopted in Nova Scotia as the official orthography. of Mi'kmaw) in square brackets following the original.

Additional resources evolve continuously. Several new on-line resources are planned by various organizations and research projects in the months and years following publication of this book, including the Pjila'si Mi'kma'ki: Mi'kmaw Place Names Digital Atlas and Website Project. We plan to regularly update a list of resources on this book's website at: www.cbupress.ca.

INTRODUCTION
THE LANGUAGE OF THIS LAND, MI'KMA'KI:
WEJI-SQALIA'TIEK—WE AROSE FROM HERE

The Language of this Land, Mi'kma'ki explores how the Mi'kmaq of Eastern North America came to mirror and express their unique relationship with the landscape they call *Mi'kma'ki*, the territory of the Mi'kmaq. This language includes the legends, songs, dances and other forms of cultural expressions—forms that mirror and communicate the rhythms and sounds, movements and patterns, and seasonal cycles of the animals, plants, winds, waterways and stars across the skies of Mi'kma'ki. These cultural expressions are perhaps the closest way we can come to understanding the biophysical character of Eastern Canada prior to colonization; they contain detailed descriptions of the behaviour, nature and characteristics of plants, animals, birds, stars, seasons and climatic changes.

The Mi'kmaw verb infinitive, *weji-sqalia'timk*[1] is a concept deeply ingrained within the Mi'kmaw language, a language that grew from within the ancient landscape of Mi'kma'ki. *Weji-sqalia'timk* expresses the Mi'kmaw understanding of the origin of its people as rooted in the landscape of Eastern North America. The "we exclusive" form, *weji-sqalia'tiek*, means "we sprouted from"[2] much like a plant sprouts from the earth. The Mi'kmaq sprouted or emerged from this landscape and nowhere else; their cultural memory resides here.[3]

The following account, written in the 1670s by the Jesuit missionary, Father Chrestien Le Clercq, provides an eloquent example of the notion of *weji-sqalia'tiek*. The oration was given by Mi'kmaw leaders from the Gaspé region of Quebec in response to an envoy of French gentlemen from Île Percée wishing to convince them of the advantages of living in the style of the French.

In response to the Frenchmen, "the leading Indians, who listened with great patience to everything answered":

…my brother, hast though as much ingenuity and cleverness as the Indians, who carry their houses and their wigwams with them so that they may lodge wheresoever they please, independently of any seignior whatsoever? Thou are not as bold nor as stout as we, because when thou goest on a voyage, thou canst not carry upon thy shoulders thy buildings and thy edifices. Therefore it is necessary that thou preparest as many lodgings as thou makest changes of residence, or else thou lodgest in a hired house which does not belong to thee. As for us, we find ourselves secure from all these inconveniences, and we can always say, more than thou, that we are at home everywhere because we set up our wigwams with ease wheresoever we go, and without asking permission of anybody…. I beg you not to believe that, all miserable as we seem in thine eyes, we consider ourselves nevertheless much happier than thou in this, that we are very content with what little we have, and believe also once for all, I pray, that thou deceivest thyself greatly if thou thinkest to persuade us that thy country is better than ours. For if France, as thou sayest, is a little terrestrial paradise, are thou sensible to leave it? And why abandon wives, children, relatives, and friends? (Le Clercq 1910: 102-104)

Le Clercq further recounts that the Mi'kmaw speaker finished his speech by saying that:

> …an Indian could find his living everywhere, and that he could call himself the seigneur and the sovereign of his country, because he could reside there just as freely as it please him with every kind of rights of hunting and fishing, without any anxiety, more content than a thousand times in the woods and in his wigwam…. (106)

Weji-sqalia'tiek is about the dynamic interrelationship between the Mi'kmaq and their ancestral landscape—a landscape integral to the cultural and spiritual psyche of the people and their language. As will be seen throughout this book, and as expressed in the Mi'kmaw language itself, this landscape was perceived as sentient, ever-changing and in a continual process of becoming. It was filled with various forms of wilful powers with which people related as relatives or enemies depending on their nature (Whitehead 1988: 2-3).

This way of thinking preceded the arrival of the Europeans and survived—communicated from one generation to another through the legends, songs, dances and oral histories in the language of the Mi'kmaq. Though postcolonial conditions have significantly changed the Mi'kmaw way of life in Mi'kma'ki, it has not significantly changed the Mi'kmaw sense of being, their sense of cultural continuity or their relationship with the land and its resources.

The unique landscape of Mi'kma'ki is the product of approximately 140 million years of geological history, the most recent being the Pleistocene epoch ending as the climate warmed approximately 13,000 years before the present (BP). A minor but significant period of reglaciation referred to as the Younger Dryas occurred between 10,800-10,200 years BP. Through these periods of erosion, uplift, glaciation, glacial scouring and deposits, rise and fall of sea levels, we have come to the landscape as we know it today (Mott 2011: 39; Province of Nova Scotia 1989: 62).

Mi'kmaw place names, along with legends and oral histories, attest to approximately 11,000 years of Mi'kmaw ancestral presence in Eastern North America, as evidenced in the ongoing excavation at Debert in central Nova Scotia, the earliest site of human habitation in Eastern North America.[4] As will be illustrated in subsequent chapters, the language and legends reflect the dramatic changes in the landscape during times of deglaciation, reglaciation and climatic shifts. Mi'kmaw legends and place names illustrate the extensive knowledge Mi'kmaq had of the diverse resources found throughout Mi'kma'ki, including resources needed for stone tools. These stone tools themselves are a reflection of the unique geological formations of the area. In turn, geological formations feature prominently in legends, which acted as oral maps of the area, while the dances and songs were a way to "tune in" to and mirror the rhythms, sounds and movements of the world (Sable 1996, 1998, 2006).

Geopolitical boundaries and foreign place names seen on maps today did not exist prior to voyages of exploration throughout the 15th and 16th centuries and colonization of the area beginning in the 17th century. Rather, the Mi'kmaq recognized seven "districts," still recognized today, with an eighth, Ktaqmkuk (Newfoundland) added in 1860.

In the 1880s, the Baptist missionary, Silas Rand, described these districts as follows:

> Cape Breton, which comprised one district, was looked upon as head of the whole. As marked on the "wampum belt," C.B. is at the head. To his right stretch away three districts with their chiefs, viz., Pictou, Memramcook, and Restigouche: and the same number to the left viz: Eskegāwaage, (from Canso to Halifax,) Shubenakadie, and Annapolis, which reaches to Yarmouth. These two arms of the country are named from two prominent points, viz., Cape Chignecto, and Cape Negro. (Rand 1875: 81)

Research compiled in the late-19th and early-20th century by Capuchin missionary Father Pacifique and Nova Scotia Museum curator Harry Piers, concur on the existence of these districts, with minor variations between their descriptions (Hoffman 1955: 521). Figure 1 is a map showing the districts according to Father Pacifique. Bernard Hoffman, in his PhD thesis

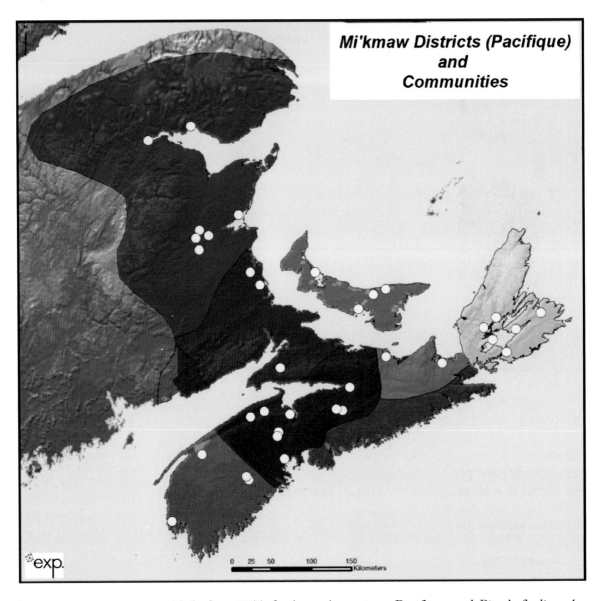

Mi'kmaw Districts (Pacifique)
and
Communities

0 25 50 100 150
Kilometers

Figure 1
Map compiled by William Jones with data adapted from Pacifique map of traditional districts found in Hoffman 1955. Sourced by exp Services Inc. Base map layers c 2012 ESRI.

published in 1955, further substantiates Pacifique and Piers's findings by correlating the districts with lists of district chiefs compiled in the mid-1700s (Hoffman 1955: 517-24).

However, Roger Lewis, archaeologist and ethnologist at the Nova Scotia Museum, theorizes that these district boundaries more likely would have followed naturally existing drainage systems, which form the principal river systems (figure 2). Each of these drainage basins is further subdivided into subunits of principal rivers and secondary streams. The myriad of rivers, streams and lakes provided a valuable resource base as well as acted as the main transportation routes for social, economic, and political interactions

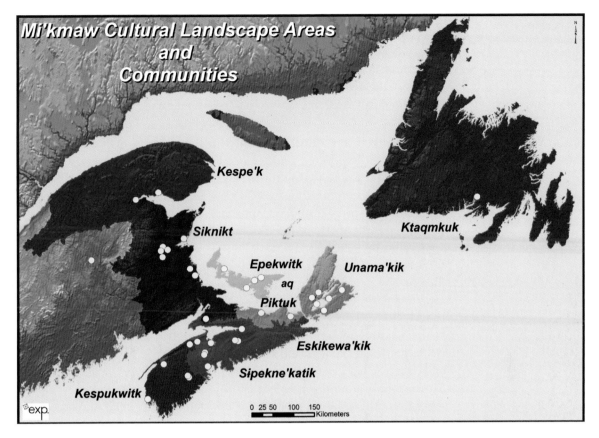

Figure 2
Map compiled by William Jones with data adapted from content contributed by Roger Lewis, Trudy Sable and Bernie Francis. Sourced by exp Services Inc. Base map layers c 2012 ESRI.

among the Mi'kmaq.[5] Furthermore, these natural boundaries were most likely flexible and permeable, reflecting changing conditions and the needs of people in each area, rather than acting as geopolitical boundaries. The following list of districts reflects Lewis's watershed theory of the districts together with the Smith/Francis orthography and current translations.[6]

Kespe'k ("end or land") is comprised of the Saint John River Valley and the Appalachian Mountain Range of northern New Brunswick and the Gaspé area of Quebec.

Epekwitk aq Piktuk (Prince Edward Island, "cradled above water" and Pictou "explosion place") is comprised of PEI and the lowland area along the Northumberland Strait, separated from neighbouring districts by the Cobequid Highlands and the Pictou and Antigonish Highlands.

Sipekne'katik/Sikipne'katik (Shubenacadie, "area of wild potato/turnip") is comprised of the Shubenacadie District and the Minas Basin Coast.

Kespukwitk ("end of flow") includes the area west of the La Have River to Yarmouth/Cape Sable in South/southwestern Nova Scotia.

Unama'kik (a variation of the word Mi'kma'kik, meaning "Mi'kmaw territory") Cape Breton Island.

Siknikt ("drainage area") includes the Miramichi River and the Acadian Coast and Bay of Fundy Region.

Eskikewa'kik (translation uncertain at this time[7]) comprised of the portion of Atlantic Coastal Region from the western portion of Nova Scotia west of Sheet Harbour to Canso.

Ktaqmkuk ("across the waves/water") Newfoundland.

Most likely, the districts outlined in the research of Rand, Pacifique, Piers and others reflected a postcolonial adaptation to the military presence of French and English vying for Mi'kmaw lands (and souls) throughout the 17th and 18th centuries. In Lewis's view, these boundaries were less about geopolitical districts than about practical and physiographic features of the landscape with which people interacted to sustain themselves physically, spiritually and culturally.

Lewis's work on the traditional-use areas of Mi'kma'ki, particularly in present-day southwest Nova Scotia, shows the importance of these drainage areas and river systems as containing a variety of ecosystems through which the Mi'kmaq moved to take advantage of animal migrations and fish runs, as well as other resources throughout the year. It is along these rivers, their rich estuaries, tributaries and surrounding landscapes that Mi'kmaq settled and created what Lewis refers to as "critical land use areas" of approximately 50 km radius around settlement areas prior to the creation of reserves in the 1800s. These areas provided a diversity of overlapping and seasonally available resources (Lewis and Sable forthcoming).

Mi'kmaq had intimate knowledge of these areas, and the variability, seasonality and availability of various plant and animal life, which were critical to their sustained existence. Lewis's in-depth research on the construction of four types of eel weirs in southwest Nova Scotia reveals just one of many examples of the ecological knowledge Mi'kmaq held of variable seasonal and physiographic conditions. Each weir construction was dependent on the migration patterns of different fish species, and the various physiographic and climatic conditions that affected their behaviour, including water levels, moon phases and temperatures (ibid.).

These same areas continue to be attractive and favoured for settlement and resource use among Mi'kmaq today. Figure 3 illustrates pre-European Mi'kmaw archaeological sites in current day Nova Scotia relative to contemporary Mi'kmaw communities. Most Mi'kmaw communities are not significantly displaced from traditional land and critical resource areas despite the disruptions of traditional lifeways (ibid.).

———

The message of this book is the language and cultural expressions that arose from living within a fluid and changing landscape, and how these expressions

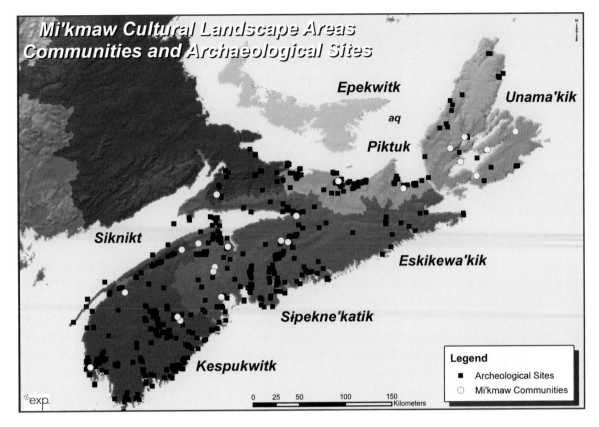

Figure 3
Map compiled by William Jones with data adapted by content contributed by Roger Lewis, Trudy Sable and Bernie Francis. Sourced by exp Services Inc. Base map layers c 2012 ESRI.

mirrored and communicated multiple layers of reality. It is within these realities that people were, and continue to be, at home—*weji-sqalia'tiek*.

The word *ko'kmanaq* is also important to introduce and contemplate. It conveys the notion of relatedness and kinship, which also carries an implicit set of values and obligations. *Ko'kmanaq* means "our relations," "our relatives" or "our people." The extended family was, and continues to be, the most important unit for the Mi'kmaq.[8] Within each of the seven districts, there might be eight to ten local communities. These close-knit communities most likely were made up of bilaterally related households.[9] These related households could be scattered along bays and mouths of rivers in such a way that the individual households could support themselves from the local resources each needed to survive. Specific resource-use areas would be determined annually by the *saqmaw* and adjusted to changing demographics and resources. At times when greater manpower was needed, such as for the fish runs (evidenced in Lewis's research in the construction of eel weirs) and game drives, people would come together cooperatively in larger groups. These gatherings would also be times of visiting, marriage arrangements and council meetings to determine important issues such as war and alliances (McGee 1977: 109).

The extension of family members is based on reciprocal relationships, whether formal or informal, and allowed—and still allows—for a young Mi'kmaw to have flexibility in residence, and go live with uncles or aunts or other relatives to learn particular skills. This flexibility and shared parenting among a number of households, gave the young man or woman wider educational opportunities, as well as a welcome home during their travels or in setting up a new residence, or an escape from an unpleasant situation (McGee, lecture November 13, 1990). Personal relationships, particularly among extended family members were and are highly valued.

Personal and reciprocal relationships extended to animals and other objects considered inanimate in Western world view, such as rocks, mountains, certain stages of the production of wood products, winds, weather and so forth, as will be seen in the chapter on legends and mentioned in the discussion of language in this book. This implies that one's relationship to the world, and its many energies and the forms that energy took, required a kind of respectful vigilance of the various forms of power, which in turn required proper conduct depending on the relationship one had with them.

In turn, the community benefited from and protected the individual. Reciprocity relied on humility, respect and generosity of spirit. Selfishness, self-proclamation or arrogance were antithetical to the community's well-being.

> When ... someone begins to assert himself and to act the Sagamore, when he does not render the tribute, when his people leave him or when others get them away from him; then ... there are reproaches and accusations as that such a one is only a half Sagamore, is newly hatched like a three-days chicken, that his crest is only beginning to appear.... (Biard 1959: 89)

Reciprocity also meant that sharing both material wealth and knowledge, was one of the basic values of Mi'kmaw culture.

> They refused nothing to one another. If one wigwam or family had not provisions enough, the neighbours supplied them, although they had only that which was necessary for themselves. And in all things it was the same. (Denys 1968: 415)

Le Clercq provides a lengthier description:

> The strong take pleasure in supporting the feeble; and those who by their hunting procure many furs, give some in charity to those who have none, either in order to pay the debts of these or to clothe them, or to obtain for them the necessaries of life. Widows and orphans receive presents, and if there is any widow unable to support her children, the old men take charge of them, and distribute and give them to the best hunters, with whom they live, neither more nor less than as if they were the actual children of the wigwam. It would be a shame, and a kind of fault worthy of eternal reproach, if it was known that an Indian, when he had provisions in abundance, did not make gift thereof to those whom he knew to be in want and in need. This is

why those who kill the first moose at the beginning of January or February, a time at which those people suffer greatly, since they have consumed all their provisions, make it a pleasure to carry some of it themselves very promptly to those who have none, even if these are a distant fifteen to twenty leagues. And, not content with this liberality, they invite these latter also, with all possible tenderness, to join their company and to remove closer to their wigwams, in order that they may be able to aid these people more conveniently in their necessity and in their pressing need, giving a thousand promises to share with them the half of their hunting. (Le Clercq 1968: 117])

This reciprocity still exists in Mi'kmaw culture. An example in Eskasoni (Eskissoqnik) is when someone dies, an auction is held called *salitte*. People donate items to the auction, then attend the auction to bid on other items, or even on those which they donated, perhaps wishing to keep it in the family. The money from the auction goes to help the family of the deceased pay for the expenses of the funeral.

Connected with sharing was the non-materialistic quality of pre-contact and early historical Mi'kmaw culture. As Nicholas Denys observed in the mid-1600s, "The hunting for the Indians in the old times was easy for them. They killed animals only in proportion as they had need of them" (Denys 1968: 426). Le Clercq also commented on how little attachment they had to the few material items in their possession as can be seen in the earlier account (Le Clercq 1968: 244).

In this book we hope to show a fluid and living landscape filled with networks of reciprocal relationships and moral obligations. We will illustrate a different conceptual framework and perception of Mi'kma'ki through a discussion of the Mi'kmaw language, and the encoded messages of the legends and the rhythms and sounds of the dances and songs. What we perceive as a literal landscape becomes a mirror of Mi'kmaw psyche, embedded in their culture, and inseparable from their being (Sable 1996, 2004). *Weji-sqalia'timk* is about an embodied landscape—a landscape that is still integral to the cultural psyche of the Mi'kmaq today, one that can still be recalled, despite increasing urbanization and institutionalization of the Mi'kmaw population.

CHAPTER 1
MI'KMAW LANGUAGE AND WORLD VIEW

Our language is verb oriented. We naturally see, and act, within the world differently than speakers of English and French, both groups speaking languages from the Indo-European stock." (Bernie Francis)

The Mi'kmaw language provides a focal point for understanding both Mi'kmaw culture and its intimate relationship with Mi'kma'ki. An Eastern Algonquian language, Mi'kmaw is the language of Eastern Canada; it grew from and is inseparable from this landscape despite cultural and social changes that have occurred since proto-historic times. Embedded in the language is a rich and extensive body of knowledge and a unique way of knowing and relating with the world in its many manifestations.

The Mi'kmaw language has evolved and diversified from Proto-Algonquian over the last 10,000 years, and became linguistically distinct from other Eastern Algonquian languages, such as Maliseet, Passamaquoddy, Abenaki and Penobscot, at least 600-700 years ago. Its linguistic differences from other associated language groups have long thwarted Algonquian scholars from making overarching generalizations regarding this language group. For instance, the Mi'kmaw words for "water" (*samuqwan*) and "stone" (*kun'tew*) should logically share common Proto-Algonquian roots with other Eastern Algonquian languages. However, these two words do not fit that mould and are unique to Mi'kmaq for reasons not yet clear.

By contemplating the nature and structure of the language, and its implicit meanings, we can gain some understanding of another world view, a different filter through which to perceive and conceive of the world than through Indo-European languages. Silas Rand, though largely unsuccessful in his efforts as a Baptist missionary to win the souls of Mi'kmaq, spent almost forty years documenting the language. He left a legacy of dictionaries and transcriptions of legends collected through the latter half of the 1800s.

Though still being corrected and revised by Mi'kmaw scholars, these documents include a voluminous and detailed terminology for the landscape and its features.

> They have studied botany from Nature's volume. They know the names of all the trees and shrubs and useful plants and roots in their country. They have studied their natures, habits, and uses. They have killed, dissected and examined all animals of North America from the *mestŭgepegajit* to the *gulwâkcheech* [from the buffalo to the mouse]. They have in a like manner examined the birds and the fish.... (Rand 1971: xi)

As an example, Rand documented more than seventy words applicable to the making of a canoe, as well as a similar number related to birchbark and its uses. According to Elders Margaret Johnson and Wilfred Prosper, both now deceased, even this was not a full accounting of the terms.

Prosper, a former chief in Eskissoqnik (Eskasoni) and a Mi'kmaw scholar in his own right, knew a litany of words associated with the different grains of wood in trees, many no longer in use. *Welipka'q*, meant that the grain ran straight and did not split off during cutting (literally, "strip or split well," according to Hewson); *ksu'skawinaq*, referred to a type of grain found on one side of a spruce. The grain of this wood was different from the rest of the tree, and was the best for making bows. Margaret Johnson also mentioned the word *kaqjet*, which means "brown, burnt-like wood" that had a hard grain and was no good for basket making. The terminology for and knowledge of the wood grain itself is extensive, not to mention the other parts of the trees. This was and is important to Mi'kmaq in terms of their numerous uses of trees.

Figure 4 (left to right)
John T. Johnson, Margaret Johnson (Dr. Granny), Regina Prosper on lap of Clara Prosper (behind, Wilfred's mother), Kenny Prosper on lap of Peter Prosper (Wilfred's father), Peter Prosper on lap of Wilfred Prosper, and Bessie Prosper. Courtesy of the Prosper family. Photographer unknown.

With the passing of Elders such as Margaret Johnson and Wilfred Prosper, many of the terms associated with the landscape are forgotten, as traditional technologies have fallen from use or land use patterns have changed. Terms for rocks and minerals, and technologies associated with their use, are difficult to find since the early introduction of iron by Europeans decreased the need for stone tools, as noted by Nicholas Denys in the mid-1600s (Denys 1983: 39). However, by studying the nature of the Mi'kmaw language along with place names, legends, songs and dances we begin to experience a living and integrated landscape and understand the comprehensive knowledge system Mi'kmaq had of Mi'kma'ki.

Language is the unique reflection and expression of how cultures structure, give meaning to, and interact with the world. Each language has its own syntax—the grammatical organization of a culture's perception and experience of reality—that best serves their needs. Each language has its own semantics, the ascription of meaning to words and word parts (morphemes) that hold the implicit values and assumptions underlying a culture's world view.

Language also has its own cadence and rhythm, a sound unique to the culture. As the Recollet missionary Abbé Maillard (Musmaya'l to the Mi'kmaq) wrote in 1755, in reference to speaking to the Mi'kmaq in their own language, "I even take care of observing measure and cadence in my words and to make choice of those expressions that properest to strike their attention, and to hinder what I say from falling to the ground" (Maillard 1758: 2). This rhythm will be further explored in discussing the songs, chants and dances of the Mi'kmaq.

Most important, is that the Mi'kmaw language is a continuous link from pre-European-contact society to contemporary Mi'kmaw culture. Many Mi'kmaq argue that their language is their culture, the loss of which would be devastating. Not only does the language continue to be vital to the culture, it is beautiful and filled with profundity. Its descriptive and flexible nature, and its ability to compress a multitude of meanings into a single word, reveals a world Western science has only begun to articulate in the last century. It reflects a world of interdependent relationships, a world in constant motion, metamorphosing and filled with the potential for new patterns, new shapes and a variety of conscious beings with whom one interacts, honours and dances or whom one conquers.

Rand wrote the following description of the Mi'kmaw language, illustrating both its flexibility and descriptiveness.

> The language of the Indians is very remarkable. One would think it must be exceedingly barren, limited in inflection, and crude; but just the reverse is the fact—it is copious, flexible and expressive. Its declension of nouns and conjugation of verbs are as regular as the Greek, and twenty times as copious. The full conjugation of one Micmac verb will fill quite a large volume; in its construction and idiom it differs widely from the English. This is why an Indian usually speaks such wretched English; he thinks in his own tongue, and speaks in ours, following the natural order of his own arrangement.... The verb is emphatically the word in Micmac. Whole sentences, and long ones

too, occur constantly, formed wholly of verbs. All adjectives of the animate gender are real verbs, and are conjugated through mood and tense, person and number..... Even the numerals are verbs, and any noun can assume the form and nature of a verb without difficulty. (Rand 1974: xxxiv, xxxvii)

One of the most distinguishing factors of the Mi'kmaw language, as noted by Rand, is that it is verb-based, not noun-based as is the English language. The verb is where everything happens; it is the focus of the language with prefixes, infixes and suffixes added to determine gender, tense, plurality, animacy and inanimacy. This focus on the verb, and the "copious" suffixes that can be added to it, allow for extraordinary breadth and creativity of expression. It makes the language adaptable, able to forge new expressions to meet life's shifting and unpredictable realities, reflecting the nature of the universe as being in a continuous state of flux, ever changing and non-static.

It also allows for great word play and humour. This style of humour (*kiso'qn*) is hard to describe, but it capitalizes on the same word having totally different meanings. An example is *so'qotemitaq*, which could mean "she/he ran up stairs crying" or "he/she vomited." A second example is *pepsi puatmn*, which means "Would you like a Pepsi?" However, the same word with a hyphen added becomes *pepsi-puatmn* and implies, "You like to have sex a lot?" Some Mi'kmaq were (and still are) masters of these double meanings, and could play with words at a dizzying speed leaving their audience with their heads spinning.[1] Some body parts also have stories associated with them, and can be a great source of humour for those who understand the meanings.

The descriptive and holophrastic quality of the language also allows for playfulness. One word in Mi'kmaw can encapsulate and create a whole picture. Aside from the inflectional endings, which indicate tense and animacy or inanimacy, each word can be made up of several morphemes, each of which has meaning. In English, it might take a whole clause to describe an image, using different parts of speech, each of which has its placement within the grammatical structure of the language. In Mi'kmaw, these parts of speech are encompassed within one word. One such word expression is *pemi'skipetesink*, "a person moving along as s/he moves his/her neck in a jerky manner." The following are the morphemes with specific meanings that when brought together in a holophrase communicate a clear image to the listener.

pem – moving along
i'skipe – neck
tes – jerky movement
in – stative
k – third person singular

Every language has its purpose, which serve particular needs. Understanding some of the differences helps us to understand the bases for some cultural misunderstandings. Within marginalized societies, in this case the Mi'kmaq, differences have repercussions that can affect well being. Even the development of Mi'kmaw orthographies (spelling systems) and translations by various missionaries, such as Chrestien Le Clercq, Abbé Maillard, Silas Rand and Father Pacifique, were politically and religiously motivated, as will be seen in a number of examples.

In the 1970s, linguists Doug Smith and Bern Francis developed a new orthography that more precisely reflected the twenty-seven or so distinctive sounds existing in modern day Mi'kmaw. During a videotaped interview with Smith and Francis in 1994, Smith had the following response to a question regarding the differences between the Mi'kmaw and the English languages:[2]

> You [speaking to Bern Francis] mentioned the verb-like quality of Mi'kmaw in that it reflects the Mi'kmaw world view where the world is perceived primarily as flow or as flux, movement as opposed to the Indo-European noun-centered languages which objectify the world; they turn the world into objects which can then be analyzed. They can be gotten hold of, taken apart, put back together and treated as things as opposed to the movements. In the Mi'kmaw language, there is an inherent dynamism or movement that Mi'kmaw speakers themselves are always aware of, whereas in English, we tend to be more aware of nouns. We are a thing-oriented society rather than a movement-oriented society. (Smith/Francis, interview September 30, 1994)

All languages shape and cut up reality differently. For Mi'kmaw speakers, however, the difference of how they perceive the world differently than English speakers can be likened to the difference between pictures from a camera and moving pictures from a video camera. A camera takes a picture of the world photograph by photograph while a video camera shoots continuous sequences of pictures without perceptible disruption.

Even the so-called nouns in Mi'kmaw are recycled verbs with a slight change to their ending causing them to be seen as a noun. Morphemes like *-ikn* and *-aqn* are suffixes, which give a verb a noun-like quality. These words can be easily shown to be old verbs. For example, the verb *ekwitk* suggests something is in the water. *Kwitn*, the word for canoe, originates from the verb infix *kwitk*. The transitive verb *taqtm* meaning "I hit it" is used in part to construct a noun, *taqtikn* or *taqtaqn* both meaning "the hitter" as in a stick for hitting a drum.

An excellent illustration of how language reflects the fluidity of Mi'kmaw world view can be seen through the changeable and varied words for the creator principle. The word *Niskam*, was adapted by the missionaries to connote the word "God."[3] There was never one word for Creator in the Mi'kmaw language, but rather a number of different verbs, mostly transitive verbs, that articulated different processes of creation. *Kisu'lkw, ankweyulkw, jikeyulkw, tekweyulkw* were all words for creator.

> *Kisu'lkw*: the one who created us; he, she, it who (or that which) created us
> *Ankweyulkw*: he, she or it who (or that which) looks after us
> *Jikeyulkw*: he, she, it who (or that which) watches after or over us.
> *Tekweyulkw*: he, she, it who (or that which) is with us

None of these words were nouns that connoted one central being as a source of creation. They are different *processes* of creation; they can refer to the creator who does all these things, or describe a role or roles in the process of creation. When these words are used, it is understood that the speaker is referring to the creator each time. As transitive verbs, these words, or any word for creator, can be conjugated more than four hundred different ways. These terms are also present tense indicative meaning that "Creator" or "God" is ongoing. You could never speak of Creator as something that has already happened, such as "When God created the world….." In other words, "God" is a process, a continuously manifesting, creative force.

The early missionaries had difficulty with the Mi'kmaw view of creation. First they were faced with a world view that experienced the universe as fluid and transforming, as was the concept of a creator. In addition, it was difficult for the missionaries to perceive God as also being potentially female or a sentient "it." In reaction, the missionaries, in need of a word to communicate the abstract notion of God, focused on the word "*niskamij*" which means "grandfather," "step-father" and "father-in-law." This is both a literal term for a kin relation, and an honorific form of address. It was also used to address the sun, who the Mi'kmaq attributed with providing the conditions to help support life and for things to grow, incorrectly interpreted by the missionaries as "impregnating the earth from which all life sprang" (Le Clerq 1968: 84; Maillard 1758: 47-48). Nailing down the concept further, the missionaries dropped the "*ij*" ending, and the word evolved to connote an unchanging, masculine God.

Not only are the words for "creator," such as *kisu'lkw,* verb-based, but they are also "we inclusive." "We inclusive" means that the speaker includes the person he/she is talking to as well as everyone else around them. The notion of God making any one person, or being personally yours, cannot exist. In this case, the "we inclusive" does not indicate whether something is animate or inanimate, male or female. The difficulty in English, of qualifying whether the Creator is a "who" or a "that"—person or object—is not present. This means that *kisu'lkw* can be male or female or a sentient object. In Mi'kmaw, an object can be animate or inanimate—the Creator could be a machine. All these possibilities are included in the word. Despite this linguistic quality, it is true that Mi'kmaq look on the Creator as sentient and therefore is a "he" or "she" or an "it."

Emerging from this discussion is another fundamental quality of Mi'kmaw world view articulated in the language—the relational, associative aspect. In Mi'kmaw, every thing or every person is spoken of in relation with something or someone else, as seen in the discussion about *niskamij,* an honorific term for the sun: *na'ku'set* (literally "the day shiner"). The moon (*tepknuset,* literally, "the night shiner") was also referred to as *kukumijinu,* meaning "our grandmother," as were certain rocks and rock formations that/who will be discussed in the next chapter. Everything existed within a network of relationships and could not exist as a separate entity outside those relationships.

A further example of this is the translation of the Catholic blessing in English referred to as making the sign of the cross, a gesture a person does on oneself, known in Mi'kmaw as *klujjiewto'simk,* which literally means, "to crucify oneself." In English, the words of this blessing are, "In the name of the Father, the Son and the Holy Ghost." The Mi'kmaw language does not have a separate word for "father." The notion of father must be attached to a possessive pronominal marker. *Nujj,* for instance means "my father," with the "n" being the possessive pronominal marker indicating "my," but no independent word exists (i.e., *ujj)* for father. Because the Mi'kmaw language does not allow for the existence of a father without a child, or a child without a father, this blessing had to be translated to reflect that relationship. In Mi'kmaw, the blessing would be, *Ta'n teluisit wekwisit Niskam,* which translated means, "In the name of the father who has a son," or *eujjit niskam* which means "in the name of the son who has a father."

In Mi'kmaw, there is also what is called the absentative case, which is used to indicate someone whose consciousness is no longer present, and can refer to a person sleeping as well as someone who is physically absent or deceased. Therefore, if a father lost his child, the absentative case would be used to designate the deceased child, and the person would no longer be a father by virtue of having no one to refer to him as such (if this was his only child). The deceased or the sleeping child would be referred to as *nijanaq,* which means "my child" in the absentative case. The ending, *-aq* indicates the absentative case.

In English, the word "father" can stand alone, separate from a concrete relationship, an abstract concept. To Mi'kmaq, the notion of a priest addressed as "father" who had no son or visible kinship ties was anomalous and abstract. How could an individual exist in isolation from some form of a relationship?

This relational quality of the language extends to the Mi'kmaw relationship to the environment. A simple demonstration of the inseparable relation between the environment and all things, mental or physical, occurs in the words for colours. Except for the four colours—red, black, white and yellow (also the colours used for the four directions)—all colours are associative or analogized. Even these four, however, are thought to have derived from Proto-Algonquian words that associate them with blood (red, *mekwe'k*), light/sunlight/dawn (yellow, *wataptek*) and white (*wape'k*) and ash (black, *maqtewe'k*). Other colour terms mean "like the sky" (blue, *musqunamu'k*), "like the fir trees" (forest green, *stoqnamu'k*) and so on (Hewson, personal communication 1996; Whitehead 1982: 71). Thus there is no way to describe the colour of blue and green rocks, or even a dream of blue and green rocks, without ascribing to them a connection, or relation, to the sky and fir trees. Furthermore, all colours—including black, red, yellow and white—are verbs. They are intransitive verbs that can be conjugated. The translation of *maqtewe'k* (black) means "in the process of being black," inferring that there is no fixed state of blackness, but rather a stage in a process that could change. The same is true for *wape'k* (white), *mekwe'k* (red), and *wataptek* (yellow). These are all verbs, not adjectives as in English, and therefore indicate being in the process of becoming white, red or yellow.

This relational aspect can also be seen in the extension of kinship terms—which we generally associate with human-human relationships—to animals, stars and other beings. All animate beings are—or have the potential for being—one's relatives and taking human shape. Therefore, kinship terms would naturally be applied. A Kluskap story recounted by Jerry Lonecloud (born Germain Bartlett Alexis) involves the use of the kin term "my uncle," *nklamuksis*, which literally means "he who looks after me." Kluskap was a legendary *kinap*, or powerful mythological hero, and features in a number of legends. In this legend, *nklamuksis* is being used in reference to a whale.[4]

> On the Island Sighignish, Glooscup's [Kluskap's] niece (they/the animals and birds/were human then) was in the woods w(ith) bow and arrow shooting small game such as squirrels, rabbits, animals for their prey, and other small animals. When she returned she found the people in the encampment had left in their canoes to go to the mainland when [Kluskap] required them. She didn't know how to get to the mainland. Finally, she saw a whale passing by. She said to the whale, "Uncle, will you be so kind as to take me to the mainland? I am here all alone." So he said, "Yes, I will take you but I can't take you on dry land. But, she said, "Well take me as near as you can." (Dennis 1923: Notebook 2. 11)

In Rand's retelling of the legend "The Two Weasels" (also in another version entitled, 'The Badger and the Star-Wives"), by choosing their favourite stars while lying awake one night, two sisters inadvertently cause the stars to transform into humans and become their husbands.[5]

Here we begin to glimpse the interchangeability of energy and shape-changing quality of the world seen throughout Mi'kmaw legends and embodied in the language. In legends, humans commonly shape-changed into animals, birds, plants, rocks and other-than human beings, and vice-versa. Birds and animals were said to come from the stars. Marriages occurred frequently between animal "persons" and humans or other beings. As will be seen later in the discussion on animacy and inanimacy, plants, rocks, mountains, thunder and many geographical features were regarded as conscious beings that could change shape unpredictably and at will (Whitehead 1988: 12).

People also had an "animal spirit helper" (*waisisl*) or a personal alliance with an animal, whom they could call upon for assistance, protection or guidance. Both their "spirits" were interchangeable and inseparable in essence. It was believed that whatever happened to one affected or transferred to the other. For example, in one legend a moose is a man's spirit helper and when the moose is injured, the man also gets injured.

This extended to community or family totems, or animals that were particularly significant to the family—meaning the extended family, involving a network of relationships that most likely formed a community as explained in the previous chapter. In one story transcribed by Silas Rand, an orphaned boy is brought up by a bear, later retrieved by his village, and from then on carried the name *muin*, which means "bear," as his family name. Because of this connection, the family was never to kill bears. The descendants are said to be the Thomas family from Pictou Landing, whose Mi'kmaw name is Muin (Wallis and Wallis 1955: 431; Rand 1971: 259). The family name Googoo comes from the word for owl, *ku'ku'kwes*. In Prince Edward Island, Snake (*Mte'skm*) was a common name. The word *ntutem* (totem) is derived from the Algonquian word *neto:te:ma* meaning "my family group." It is cognate with the Cree term *nito:te:m* meaning "my relative" or, as a term of respect, meaning "spiritually a member of my clan" (Hewson, personal communication 1996).

On all levels of reality—visible and invisible, everything was related and thus required that a person act with proper decorum. One could not take these relationships for granted, but needed to respect and honour them properly, including the negative or hostile powers. For instance, in February, the harshest time of the year, food was placed outside for Apiknajit (literally "snow blinder") to show respect for the fierceness of the month of February and as a way to protect members of the community from hunger (Maillard 1755: 23). It is this type of offering that embodies the relationship to the seasons, the earth and the universe.

This offering is still practised today by some Elders such as Annie Prosper of Apaqtnekek (Afton), Nova Scotia. In fact all of the months in Mi'kmaw are animate, each regarded as having a special responsibility for care of the Mi'kmaq, such as January, which is the month that provides the tomcod. Apiknajit is the only month that does not end in the suffix *-iku's*, which refers to the moon. It could be speculated that it was very important to feed the entity, Apiknajit, as a reminder to pay to attention and show respect, and not take anything for granted.

Offerings, gift-giving and proper conduct were ways people fulfilled their obligations to others for the gifts they were given. Gift-giving was an inherent part of Mi'kmaw culture, whether in the form of dance, song, feasts or material presents. No chief visited another without bringing gifts (Maillard 1758: 4). They in turn would be honoured by feasts, dances, orations and many other forms of offering. If one person harmed or insulted another, they would atone by giving gifts—a way to acknowledge relationships and maintain the proper balance within the many levels of existence. Life was a continuous and fluid reciprocal exchange.

Another pertinent and final example of the relational quality of Mi'kmaw world view is the Mi'kmaw word *mkamlamun*, "the heart."[6] In this word, the intuitive and intellectual, or the cognitive and emotive, are inseparable. The whole concept embodied in this word is better translated "heart/mind." When a Mi'kmaw speaks, it is from the notion of mind and heart being inseparable, and not simply that the mind is a function of the brain. A Mi'kmaw speaker might say, "*Aq teluey ni'n nkamlamunk*," or "And then I say in my heart..." meaning heart/mind.

As we just saw, the notion of mind is implicit in the word for heart—you cannot have mind without heart. The question arises whether anything or anybody can be extracted and isolated from any relationship or constellation of relationships, in which he/she/it exists (e.g., can one be a father without a son?).

How a man or woman perceives his/her "place," or finds identity, in relationship to their environment or the universe in general, varies within different cultures. However, the nature of this relationship affects how people relate with their world, and how they position themselves in social interactions. It influences how people address one another, use and view their bodies and body parts in various social contexts, and relate to their environment.

In Mi'kmaw, the relationship is indicated by the inflectional ending at the end of the verb. Although the Mi'kmaw language has a pronominal system, pronouns would generally be used only to add emphasis. For instance, if we were discussing seeing a mountain, the phrase in Mi'kmaw would be *nemitu kmtn*. Even though *nemitu kmtn* doesn't include the first-person perspective "I," it is understood that "I" am the one seeing the mountain, by inflection. *Nemitu* means "I see it," and is the transitive form of the verb indicating a transition between the subject and object with "it" being the focus of attention—the focus is on the object not the subject. So, in the phrase, *nemitu kmtn*, the focus is on the mountain (*kmtn*) being seen, giving a greater degree of importance to the mountain than the "I" seeing it. This principle is true with all transitive verbs. Whereas if we used the form, *nemitmk* the subject is indefinite and there is no predicting who is seeing the mountain; it's just saying "the mountain is seen." All these possibilities happen in that one word.

This discussion centres around Mi'kmaw being verb-based; the emphasis is on the verbs. A subtlety within the Mi'kmaw language is that the relationship is already acknowledged as present and ongoing. The subject and object are merely tacked onto this already-existing relationship; they are inferred through inflectional endings. In English, by comparison, the subject and object are established before the relationship (the verb or verb phrase) is added to bring the two into contact.

This points to a subtle difference between Mi'kmaw and most Western languages: the placement of the self in the language structure is not the central feature. In fact, there is no distinct, separate word for self. It is only inferred by the inflectional ending added to the verb implying that the self is part of a web of ever-changing relationships. The structure of the language indicates that a Mi'kmaw does not put him or herself in the forefront of anything: they seem to leave themselves second to other things or other people. The focus will be on another individual first and then the speaker will be second.

This deferential quality is deeply ingrained in the culture, evidenced in the selection of leaders. *Nikanus* is the word the Mi'kmaq used for their head spokesperson, while the term "chief" was later adapted by the colonial governments and now enshrined in the Indian Act. *Nikanus* would be chosen to lead a particular community because he would put him/herself out front as a protector and a provider, using their skills, strength, knowledge and wisdom to protect the community. They were chosen because they were the best providers and the best protectors rather than as a result of having accumulated a lot of wealth, which in the pre-contact and early post contact periods might have been food, furs, or even bodily strength. As mentioned in the Introduction, the opposite of these traits would be *newtite'lsit*, or someone who thinks only of oneself or had no consideration for others, a trait frowned upon.

The topic of personal identity or how a person perceives their place in the world has powerful implications for everyone and has profound repercussions for how we relate with our environment and other people throughout our lives.

PAST, PRESENT AND FUTURE TENSES IN MI'KMAW

Mi'kmaw has no word for time. Storytelling as it is done in Mi'kmaw, has caused difficulties for Mi'kmaw-speaking (and thinking) students in writing

English essays because they do not follow Western logic for sequencing time. Instead, they often tell stories in the present tense, as though something that happened long ago is happening now. Consequently, despite writing what they consider a good story, they often get lower grades in conventional non-native education programs. Invariably they are told by their teachers that they did not have proper grammar, and their logic needed sequencing. One Mi'kmaw woman reported being advised by an Elder that she would do fine if she just learned to think in a linear fashion.

When Mi'kmaw Elders tell a story, they seem to spiral inwards from the general to a specific point, then out again, as though first creating a landscape out of which something arises. They may also begin at any point in a story, not conscious of chronological time. This can be frustrating for researchers at first who want to "get to the point" and are seeking an answer from an Elder to a specific question. Instead, the researcher might be treated to a tale or recollection of a dream that might initially appear unrelated until, unexpectedly, the answer arises from within the story.[7]

Similarly, Mi'kmaw legends do not necessarily have conclusive endings, which is what may have prompted the Baptist missionary Silas Rand in his rendering of Mi'kmaw legends, to add finalizing sentences to the end of some of the stories. For example, Rand wrote at the end of "The Magic Dancing Doll": "Here the story abruptly ends. One feels strongly inclined to supply what may be supposed a 'missing page'..." (Rand 1971: 13). The stories themselves do not always follow a logical sequence, but the emphasis is more on the person's actions, not whether the sequence of events displays continuity or logic (McGee, interview December 1995).

Bringing the past into present consciousness can be seen in the feast orations. Feast orations were an inherent part of traditional Mi'kmaw events honouring the host or an esteemed guest. During these orations, Mi'kmaq might regale the host for hours with exaggerated praise of the honoured one's ancestry, and extolling the virtues their ancestors had bestowed on them. Similarly, at powwows, people speak about the ancestors being present, as though all levels of the apparent and unapparent phenomena are always already present in each moment. At spirit feasts, Mi'kmaw traditionalists, or Mi'kmaq who have chosen to practise Native spirituality, say you are eating for all the ancestors when you eat. In sweat lodges, pipe ceremonies and smudges, the ancestors are invited or invoked into the circle. They are present, as though time collapses into a plane on which past, present and future co-exist or do not exist at all.

The Mi'kmaw language does have a simple past tense, a future tense, a reported past, an "if conjunct" past tense, and an "if conjunct" future tense. The simple past is used most often when it is necessary or important to be specific, or to use as emphasis for a point. The past time is often for a consciousness not presently there, or to deliberately distance oneself (i.e., "I was drunk, but I am no longer)." As illustration:

teluisi: my name is _____
teluisiap: simple past – my name was _____
tluisites: future – My name will be _____
teluisiass: reported past – it is reported that my name was _____
tluisiasn: if conjunct past – if my name were _____
tluisikk: if conjunct-future – my name would be _____ if…
teluisianek: when conjunct past – when my name was _____
tluisian: when conjunct future – when my name is _____

In his account of the Mi'kmaq of the Gaspé, Quebec, the French missionary, Chretien Le Clercq, describes the Mi'kmaw concept of time as follows:

> They count the years by the winters, the months by the moons, the days by the nights, the hours of the morning in proportion as the sun advances into its meridian, and the hours of the afternoon according as it declines and approaches its setting. They give thirty days to all moons, and regulate the year by certain natural observations which they make upon the course of the sun and the seasons. They say that spring has come when the leaves begin to sprout, when the wild geese appear, when the fawns of moose attain to a certain size in the bellies of their mothers, and when the seals bear their young. They recognise that it is summer when the salmon run up the rivers, and when the wild geese shed their plumage. They recognise that it is the season of autumn when the waterfowl return from the north to the south. As for the winter, they mark its approach by the time when the cold becomes intense, when the snows are abundant upon the ground, and when the bears retire to the hollow of the trees, from which they do not come forth until the spring…. They have no regular weeks; if they make any such division it is by the first and second quarter, the full and the wane of the moon. All their months have very expressive names. They begin the year with the Autumn, which they call *Tkoors*; this expresses that the rivers begin to freeze, and is properly the month of November. *Bonodemeguiche* [ed.: this is an error and the word should be *pnatmuiku's* and is the month April] which is that of December, signifies that the Tomcod ascends into the rivers [ed.: December should be *kesikewiku's* which means winter month moon].... And that is the way with the other months, each of which has its particular designation. (LeClercq 1968: 137-39)

Time was reckoned by natural cycles of birth and death, the waxing and waning of the moon, animal migrations and the seasons. Survival, it seems, depended on understanding these natural rhythms. Stories, songs and dances, mirrored these rhythms such as in hunting dances that mimicked the movement of animals (136). It could be said that Mi'kmaq were more aware of time than cultures that abide by fabricated "clock" time.

This sense of cyclical time can be seen in the preparation for a hunt or tanning a hide, and how a person relates to this undertaking. A person in a reciprocal relationship with the rest of the universe honours harmony and cooperation, avoiding disaster or death. A hunter, for instance, would prepare for the hunt through dances and chants. The Mi'kmaq considered the animal to have its own will, and requested that it give its life. Once the animal gave it, the women then went to butcher and bring the animal back to camp, singing and dancing as they went. Then a feast of thanksgiving might be held, with more songs and dances. The animal was divided up according to protocol and need. Bones of the animals were placed in appropriate places (bones were never left on the ground) so that the animal could reanimate and return to its habitat, and continue to care for the people. The cycle continues. (McGee lecture November 6, 1990; Le Clercq 1968: 118-19).

ANIMACY AND INANIMACY
(SENTIENCE AND INSENTIENCE)

Another major characteristic of the Mi'kmaw language is the distinction between things which are animate and those which are inanimate. This distinction can be considered in two lights: syntactically and semantically—the latter a continuing source of debate for scholars. From the purely grammatical point of view (syntactically) all nouns in the Mi'kmaw language, according to Doug Smith, can be put in one "basket" or the other—animate or inanimate. Animacy and inanimacy are distinct grammatical categories, with animate words designated by specific suffixes and a particular plural ending. Noun endings must also agree with the verbs in terms of animacy, inanimacy and plurality (Smith/Francis interview September 30, 1994).

From the grammatical point of view, the largest and most consistent category of animate beings are humans and "zoologically alive entities," with other noun categories, such as flora, being inconsistent, or following no set rule for animacy or inanimacy (ibid.). For instance, strawberry (*atuomkmin*) and raspberry (*klitaw*) are animate, while blueberry (*pkwiman)* is inanimate. Tobacco (*tmawey)* is inanimate, as are apples, but potatoes are animate (ibid.).

At the level of semantics, the question of animacy and inanimacy takes on a deeper significance in trying to determine world view. What criteria are at work in the mind of the Mi'kmaw speaker to intuitively, perhaps unconsciously, designate an object as inanimate or animate? This also is very fluid, even today as new material objects are introduced into the culture, some Mi'kmaq regard them as animate and others as inanimate. John Hewson stated, "It is obvious that animacy has to do with life.... What is at stake in Alg[onquian] is clearly 'manitou power'—a spiritual element. Even if we know this much, the details have not always been worked out" (Hewson, personal communication 1995).

There are, however, some interesting insights. Animacy may have to do with the level of importance an object might have in one's life and survival, or to the working of a mechanism. For instance, a town bus on the Membertou Reserve in Sydney is inanimate, whereas on the Eskasoni Reserve it is animate, perhaps because Eskasoni, being further from Sydney, is more reliant on the bus for transportation. The frame around a door is inanimate but the door itself is animate. A television as a composite of frame, tube and wires is inanimate, but when the outside box/frame is discarded, the picture tube itself is animate, possibly because it is what is necessary for images to appear on the television. "Refrigerator" is animate, possibly because of its necessity in preserving food. However, the logic breaks down by Francis's own admission. Cars in both the Eskasoni and Membertou are inanimate, whereas a motorcycle in Membertou is inanimate but animate in Eskasoni. This variation is strange since these two communities are only fifty kilometres apart.

An object can also change from animacy to inanimacy depending on its function or use. The word *kmu'j* (tree) is animate, but can equally refer to a stick lying on the ground, which is inanimate. If the stick were made into a bow, the entire bow including the stick would be animate. As another example, if plasticine (inanimate) is moulded into a human figure, it becomes animate. Nicholas Denys, the 17th-century French merchant and explorer, described a Mi'kmaw burial in which a copper pot was placed in a grave along with other items, and punctured with holes. These goods were to serve the deceased in other realms of existence. Denys, in an effort to dissuade the Mi'kmaq from this belief, tried to argue nothing had travelled to the other world. One of the Mi'kmaw replied:

> "Do you not indeed see," said he, rapping again upon the kettle, "that it has no longer any sound, and that it no longer says a word, because its spirit has abandoned it to go to be of use in the other world to the dead man to whom we have given it"? (Denys 1968: 440)

Similarly, a mountain is sometimes animate and sometimes inanimate. Rivers, lakes and streams are inanimate, but stars, sun, moon and thunder are animate. The verb form for the word thunder, kaqtukowik, is implied by the Elders to mean "the ancestors are raising their voices." In the story "They Fetch Summer," recorded by Elsie Clews Parsons, three brothers go to ask "Sky," who is seen in human form, for warm weather. There is a type of sleet (*msikn*) that forms a blanket of ice over the branches of trees that crackles when it moves. This too is animate, whereas other types of ice may be inanimate (Smith/Francis interview September 30, 1994; Parsons 1925: 74).

The question of animacy and inanimacy is still under debate among linguists, and one that requires far more research. Whatever the underlying psychological basis for Mi'kmaq attributing animacy to certain objects, we know the flexibility of the Mi'kmaw language allows for the articulation of a world filled with the potential for things to metamorphose and manifest animate properties. It is a world of social relations, with objects and "persons" that are not necessarily of human or animal form, or have attributes familiar to non-Native thinkers. This is mirrored in the legends, where animals, rocks, people, plants, stars, thunder, lightning and a pantheon of other-than-human beings, shape-change—unpredictably and at will. After conducting numerous interviews with Elders throughout Nova Scotia, Ruth Whitehead also came to this beautifully articulated realization:

> To the Old Ones of the People, Creation itself was fluid, in a continuous state of transformation. Reality was not rigid, set forever into form. Here form changed shape according to the will and whim of the Persons manifesting those forms, at any given moment. This Creation is clearly depicted in Micmac stories: not only through their content, interestingly, but through their basic structure and the language in which they were told.
>
> Modern science maintains that all matter is energy, shaping itself to particular patterns. The Old Ones of the People, took this a step further: they maintained that patterns of Power could be conscious, manifesting within the worlds by acts of will. They thought of such entities as Persons, with whom one could have a relationship. (Whitehead 1988: 2-3)

The Mi'kmaw language is highly descriptive, creative and playful, and can encapsulate multiple layers of meaning within one word. Through the language, we perceive the world as a web of relationships, including with "objects" that many in the Western world might term inanimate or non-sentient. An island or rock becomes Kluskap's canoe; an inanimate stick becomes part of an animate bow. The flexibility of the language allows these semantic shifts to occur.

CHAPTER 2
THE SENTIENT LANDSCAPE
AND THE LANGUAGE OF THE LAND

"The land is always stalking people. The land makes people live right. The land looks after us. The land looks after people." From a Cibecue Apache quote collected by Keith Basso (Basso 2000).

Mi'kmaw legends and creation stories abound with information regarding the various landscape formations throughout this territory. Numerous Mi'kmaw legends tell of animals and people being transformed, or transforming themselves, into stones, trees, mountains and islands. Mi'kmaw place names also tell of features of the landscape, historical events and important resources, and acted as a mnemonic device to help people find their way. In fact, the mapping skills of the Mi'kmaq and their intimate knowledge of the landscape were highly respected by foreigners travelling through Mi'kma'ki.

> The Indian has studied Geography…. And most especially does the Micmac know about Nova Scotia and the places adjacent. Show him a map of these places, and explain to him that it is "a picture of the country," and although it may be the first time he has even seen a map, he can go round it, and point out the different places with the utmost care. He is acquainted with every spot. He is in the habit of making rude drawings of places for the direction of others. One party can thus inform another at what spot in the woods they are to be found. At the place where they turn off the main road, a piece of bark is left, with the contemplated route sketched upon it. The party following examine the *luskun* [ed., *lu'kwaqn*: "the map or the pointer," in the sense of the direction it points] as they term it, when they come up, and then follow on without any difficulty. (Rand 1850: 25)

Rand continues to describe the Mi'kmaw sense of distance and direction:

And "here" said the tawny guide, who was years ago directing a party in their travel from Nictaux to Liverpool in the winter, "here just half-way." When the road was afterwards measured it was found that the Indian was correct. (Rand 1850: 25)

The 17th-century missionary Father Chrestien Le Clercq and M. Dièreville, a French traveller to Nova Scotia in 1699, made similar observations, also noting the use of wampum, sticks and simple drawings to convey extensive information (Le Clercq 1968: 136; Hoffman 1955: 235). It was this use of simple marks and drawings as mnemonic devices that lead Le Clercq to develop a system of hieroglyphics with which he taught the Mi'kmaq lessons in Christianity.[1]

> ...they have much ingenuity in drawing upon bark a kind of map which marks exactly all the rivers and streams of a country of which they wish to make a representation. They mark all the places thereon exactly and so well that they make use of them successfully, and an Indian who possesses one makes long voyages without going astray.... (Le Clercq 1968: 136)

GRANDMOTHER AND GRANDFATHER ROCKS

Prominent and/or anomalous landscape features, such as large rocks or rock formations, also acted as guide posts for Mi'kmaq travelling along the waterways of Eastern Canada much like latter day light houses. These rock formations were honorifically referred to as Kukumijinu ("our Grandmother") or Kniskamijinu ("our Grandfather") and commonly have legends associated with them, such as of Kluskap turning his Grandmother or Grandfather to stone. Mi'kmaq travelling by these sites would offer punk,[2] tobacco or other offerings in respect and supplication, as if to an honoured relative. The following was related by Jerry Lonecloud to travel writer Clarissa Archibald Dennis:

> ...one [of the grandmothers is] on Carleton lake (runs into Tusket river) west branch of Tusket) on the east side of Carleton lake. There are a lot of islands there & Googmeginwayinook [Kukumijinu] is a rock there of granite. All these are granite. She faces north here.

> When the Indians visited this rock in their canoes they always left something on the rock for *googmeginwayinook*. Indians always leave something—piece of meat or tobacco or punk. She was a great smoker. Punk is grown on a tree bulges out like a knot. It is very light and when the Indians would strike rocks to get a spark they would have the spark fly on the punk and it would burn and they always presented her w(ith) tobacco & punk to [*sic*; smoke]. [Clarissa Archibald Dennis 1923: MG1 Vol 2867 #1 Field Notebook 1]

Lonecloud described another Grandmother as having a blanket over her shoulders, and a table by her side (Dennis 1923: Notebook 1: 4).

Elsie Clews Parsons, an American Anthropologist writing in the early 20th century, documented a Grandmother near Baddeck (Apatakwitk: "reversing flow") in the account entitled "Gluskap's Grandmother": "According to Mrs. Poulet, she was Bear Woman, and at Grandmother Mountain (*gomijagune'wu*) near Baddeck, she is turned to stone.... Here is a stone which predicts storm. It gets wet before storm"[3] (Parsons 1925: 86).

Ruth and Wilson Wallis documented a "Grandfather" near Upper Musquodoboit, Nova Scotia that resembles "a recumbent figure of a sleeping man covered by a blanket."

> Micmac call it "grandfather," [Kniskamijinu] and say it is an old Indian who went there to hunt. After he lay down to sleep, he was transformed into the stone, as he is seen today. In 1912 the older Indians laid a penny or some small offering by it and "made their wishes on it," in the expectation of fulfillment at this spot. My Micmac guide did not specifically name Gluskap. (Wallis and Wallis 1955: 154)

Interestingly, the term for rock, *kun'tew,* is an inanimate noun. When the question is posed how something inanimate could be viewed as a "Grandmother" or "Grandfather," one sees that it is not the outward image alone but the *experience* of the rock that brings it into the animate realm. In the case of Grandmother (and Grandfather) "rocks," the rock is no longer a rock but becomes a Grandmother because it is experienced that way. Similarly, if a rock is shaped like a bear it might become animate in the mind of the perceiver and referred to as a bear, ceasing to be a rock. Because the rock is now experienced as a bear or bear-like, a person would relate to it as a conscious being and therefore it is animate.[4]

These terms are "we inclusive," which is significant because it means that they are "Grandmothers" and "Grandfathers" to anyone who comes within the boundaries of the Mi'kmaw world. A tremendous amount of respect was and is shown to these rocks because they are considered animate, and perceived as Kukumijinu ("Our Grandmother") and Kniskamijinu ("Our Grandfather"). Aside from having human-like features and being perceived as conscious beings, they were given honorific titles because the role they played for Mi'kmaq was *like* that of a Grandmother or Grandfather. Like Grandmothers and Grandfathers, they kept an eye on people and protected them by helping them find their way. Once the perception of the rock as a Grandfather or Grandmother occurs in the mind of the perceiver, one feels comforted and not alone.

As an example, in Laurie [Provincial] Park by Grand Lake, Nova Scotia (Kji-qospemk), there is a Grandmother site documented by Clarissa Archibald Dennis from her interviews with Jerry Lonecloud (figures 5, 6).[5] It is a prominent outcrop that archaeologist Roger Lewis explained would be about the half-way mark for people travelling along the Shubenacadie River from Halifax Harbour (Kjipuktuk – Chebucto: "at the great harbour") to Shubenacadie, Nova Scotia. The outcrop offered travellers a number of desirable features—it was prominent and therefore easy to see; it was an easy place to land as well as get shelter; and if a person climbed to the top of the rock outcrop it afforded a commanding view of the lake, so any visitors or invaders could be spotted, welcomed or attacked, if necessary.

According to Lewis, the Shubenacadie River was a major travel route and water system. There are dozens of archaeological sites on the north side and mouth of the river dating from the middle-Archaic to the Woodland period attesting to continuous use of the area. Across the lake from the Grandmother are fifty acres of reserve land owned by the Sikɨpne'katik First Nation.

As will be seen in the discussion on legends as maps in the following chapter, these sites were often woven into legends and associated with specific resource areas (Sable 1996: 210-18; 2006: 170-71; 2011: 157-72). For example, in Silas Rand's transcription of "Wizard Carries Off Glooscap's Housekeeper in Micmac by Thomas Boonis of Cumberland, June 10, 1870" (Rand

Figures 5, 6
Grandmother rock on Grand Lake (Laurie [Provincial] Park). Photos by Joseph Szostak from the collection of Trudy Sable.

Figure 7 (bottom) Grandmother Rock on Tusket River. Photo taken in the 1990s. Photo by Trudy Sable.

Figure 8
Cape Split. Photo by
David Sable.

1971: 287), we learn of a variety of rock formations taking on characteristics associated with daily activities and relationships:

> Glooscap goes from Partridge Island to Cape Blomidon [Metoqwatkek: "bushes extending down the bank"; also Tkoqnji'j: translation unknown], where he decks his aged mother out in beautiful minerals, goes back to Spenser's Island where he butchers cooks and eats animals, turns his kettle upside down to form an island called Ooteomŭl [Wtuoml: "his/her pot"] then to the River Hebert, and over to pitch his tent near Cape D'Or [L'mu'juiktuk: "place of the dogs"]. (Rand 1971: 291-93)

The area around Blomidon is rich in amethyst, as well as other minerals such as the catoyant (like a cat's eye at night) faerolite. Blomidon itself is a very prominent feature in the Minas Basin and is associated with legends about the creation of the Minas Basin. In a legend about "Toad," documented by Dennis in her interviews with Lonecloud, Kluskap turns Toad into a Grandmother, who sits at Cape Split, out of respect for her work cleaning his wigwam [*wikuom*] and inspecting Kluskap's food.

> Then Glooscup got some meat she always inspected the food she was always a great inspector of food—to see there was no vermin or insects in it before it was served to Glooscup. Her reward from Glooscup was she was turned to a stone. Googmeginuwahnayinook (meaning our great Grandmother) Glooscup's grandmother sits @ Cape Slip [*sic*; Cape Split] looking out. Glooscup turned her into a stone to be remembered & seen by different tribes & he promised G [oogmeginuwahnayinook] that [?] she should be turned to a person when he returned. So she looks in that direction ever since expecting him to return. (Dennis 1923: Field Notebook 1, 8)

Figures 9, 10
Kluskap Cave at
Kelly's Mountain.
Photo collection of the
Nova Scotia Museum.

Looking at the accompanying picture of the cliff at Cape Split Nova Scotia (figure 8), one can see a profile of a face in the shadow being cast by the cliff. For some, this might be a trick of light, but for others, this may be *Kukumijinu* looking to where Kluskap departed and awaiting his return.[6]

Along with Grandmothers and Grandfathers, there are a number of caves throughout the Maritimes that are associated with Kluskap or other legends. Kluskap is also referred to as Grandfather in some cases, though this is uncommon today. One of the most visited and well known is at Kelly's Mountain (Mukla'qatik) in Cape Breton where Mi'kmaq continue to go to fast and do ceremonies (figures 9, 10).

In the early 1900s, Elsie Clews Parsons transcribed the following account entitled "Seven Men Visit Gluskap" while at Chapel Island (Potlotek), Cape Breton. In a footnote where she cites Frank Speck (Speck 1915), Parsons notes that her informant, Isabelle Googoo Morris, heard the story from her grandfather, Peter Newell. Mrs. Morris speculated that the visit was made two hundred years ago dating it to the early 1700s, if accurate (Parsons 1925: 87).

There was seven men who went to try and see Gluskap at Smoker [Saqpi'ti'jk: "at the misty place"]. My grandfather's father was with them. They took one hundred bark torches. They went through the big stone door. They walked and walked, until they burned fifty torches. [Footnote 4: Here there was a discussion between Mr. and Mrs. Morris about the actual distance covered, one mile they concluded.] They could not find Gluskap so they agreed to come back. Gluskap was a witch. The stone door was open when they went in, it was more than half closed when they came back. They could just get through.

Ever since then, nobody has tried to get in to see Gluskap. It is not worth while to go in, because Gluskap is a witch.

One time a man said he was going to go in. That stone door was wide open then. It was a stormy night, he didn't know where he was going to stay. He found Gluskap and said, "Dear Grandfather, can I stay with you till morning?" In there was a lovely wigwam. Gluskap's grandmother was there. She was glad to see the man. She said, "You are the first stranger we have seen in this world. You can stay till morning." The old lady put on a bark pot and cooked some moose meat. In the morning his socks were all nearly dry. Gluskap went out with him, went half way, then the man came out of his boat. The man's name was Joe Nuelich (little Newell). (Parsons 1925: 87)

Offerings were also left outside these caves, similar to the Grandmothers as seen in Parsons account, "Offerings to Gluskap."

Gluskap's door is at St. Ann's [Parsons 1925: 87, n. 1: "See Speck 5: 146, 156"]. There you would throw in some dry punk and a little fish for his fire and food. You say, "I wish you give me good luck." Gluskap does not want anybody to come inside. "If anybody wants anything he can put something for Gluskap outside on stone." (Mrs. Poulet)

When you go to see Gluskap, at Smoker, Cape North [Saqpi'ti'jk: "at the misty place"], you say, "My dear grandfather, I just come on your door. I want you to help me." You leave money inside door, piece of silver. [Footnote 2: Mrs. Morris was very positive that nowhere else are presents ever made to Gluskap.] You take two or three stones away with you, that's your luck." (Mrs. Morris)

At Cape Dolphin [sic Cape Dauphin, Kukumijinaq: "at our Grandmother's"], Big Bras d'Or [Pitu'pok: "long dish of salt water"], there is a door through the cliff, Gluskap's door. Outside, there is a stone like a table. Indians going hunting will leave on it tobacco and eels, to give them good luck. They do this today. (Stephen Nevin in Parsons 1925: 87)

Caves are often inhabited by *wiklatmu'jk* (singular is *wiklatmu'j* and alternatively called *puklatmu'jk*: "small people") such as at Kelly's Mountain (Mukla'qatik) in Cape Breton and in Kejimkujik ("the place of the fairies") in southwestern Nova Scotia. In Eel Ground, New Brunswick, there are caves beside the river where *puklatmu'jk* are said to live. People respect the caves and are careful to not intrude on the *puklatmu'jk*.[7]

Parson's recorded this story associated with Salt Mountain (Wi'sikk: "shape of a beaver's den") in the We'koqma'q (Whycocomagh) First Nation, Cape Breton about *wiklatmu'jk*.

> These supernatural beings are about four feet tall. They can see at a distance, "as far as from here (Whycocomagh Reservation) to Salt Mountain." When she was a little girl, Isabelle Googoo saw one out on the bay one day in a dory with her grandfather. She saw him on Salt Mountain. "look at that little fellow climbing on the stone, on the rock wall," I said to grandfather. "Don't mind him, that's a witch," said grandfather. (Parsons 1925: 96)

A similar story was recorded by Ruth and Wilson Wallis about *puklatmu'jk* at Tracadesh Mountain on the Gaspé Peninsula in Quebec, and in Cape Breton. They are described as:

> …a race who live in mountain caves. They are small people, but they do big work. They live and dress like old time Indians, and eat only wild meat…. Like other supernatural beings they can help and they can harm. They can give a man furs, or warn of coming evil, and they can perform tricks around the house and barn. (Wallis and Wallis 1955: 154)

Wiklatmu'jk, as they are known in Nova Scotia, are generally regarded as mischievous rather than being malevolent. For instance, *wiklatmu'jk* might go into a barn and tie horses tails together or stuff the chimney with something to make it smoky (ibid.). Some people say they are evil and did things like knock over Christian crosses when the settlers came, or sit on your bedpost and laugh at you. Others consider them nice and that say they will help you out of the woods if you are lost.

The *wiklatmu'jk* were and continue to be an important part of the education of the Mi'kmaw people. They are revered (and feared) because of their crucial role in helping to adjust the mindset of the Mi'kmaq toward respectful and sensible behaviour with other members of the tribe. They do this by upsetting things when "the golden rule," so to speak, is violated, which causes great consternation among the Mi'kmaw people. Fear arises in the mind of a Mi'kmaw when he or she knows that they have done something underhanded or disrespectful to a fellow community member. That fear arises because that person does not know what tricks the *wiklatmu'jk* may be planning to restore the respect and balance that has been violated.

Wiklatmu'jk are still part of the Mi'kmaw cultural psyche, whatever the experience of them or feelings toward them. They are associated with the landscape today—landscape features throughout Mi'kma'ki—and are yet another energy or power to relate to and respect.

PLACE NAMES

Along with the numerous "relatives" associated with landscape features, such as the Grandmothers and Grandfathers, place names in Mi'kmaw also tell a story of the land. Place names not only tell of features of the landscape, historical events and important resources, but act as a mnemonic device to remind people of how to "live right." Thomas Andrews, in conducting research into the significance of place-names in the Dene culture, stated:

> Place names provide a "hook" on which to structure the body of narratives, and in doing so, become an integral part of the narrative itself. This is particularly evident in myths and legends recounting the travels and exploits of mythical heroes, which list in great numbers places relevant to the story line. Place names are therefore mnemonic devices, providing a mental framework in which to remember relevant aspects of cultural knowledge.... It is clear … that within many societies possessing rich oral traditions, landscape may be viewed as a collection of symbols which record local knowledge and meaning, and where place-names become memory aids for recalling the relevance of a "message" encoded in associated narratives. Physical geography is transformed into "social geography" where culture and landscape are fused into a semiotic whole. In essence, one cannot exist without the other. (Andrews 1990: 3, 8)

Andrews cites a number of examples from various cultures, such as the song lines of the Australian Aborigines, the Roti of Indonesia and the work of Keith Basso among the Cibecue Apaches, all of which demonstrate the inextricable union between these cultures and the land. Similarly, in his own research with the Dene of the Mackenzie Valley in the Northwest Territories, Andrews found particular stories were told at specific locations. For example, at two specific locations, moral stories were told regarding the consequences of excessive gambling. As the stories unfolded, two members of the party retreated, not joining in the humour and camaraderie of the story. These men had expressed an interest in gambling whenever they reached their destination (Andrews 1990: 17).

During research into traditional land use practices of the Mi'kmaq and Maliseet for the Canadian Parks Service (CPS), several Mi'kmaq suggested that place names be integrated into any interpretation of the land. It also became apparent during this research that, while CPS was looking for "in situ," visible evidence of land use within park boundaries, the Mi'kmaq and Maliseet placed much more emphasis on the oral traditions associated with various regions. For instance, at that time Fundy National Park had produced little archaeological evidence of First Nations' presence in the area, despite the fact that a strong oral tradition exists regarding the use of the land, valid evidence of presence to the Mi'kmaq themselves (Sable 1992; Joe Knockwood, interview 1991).

The notion that narrative tradition is rich in place-names that are mnemonic devices providing a framework by which to remember relevant aspects of cultural knowledge, does not seem so different from European cultures, and this fact alone would not justify the claim that traditional Mi'kmaq held, and still have, a different world view. It is the *power* of the stories and the consequent significance of the place names to individuals within the cultural community that gives us a glimpse into what can be termed another world view. The Mi'kmaw culture, essentially, is inseparable from the land of the Eastern Canada.

Andrews cites a number of semantic categories devised by Cornelius Osgood in his research into Gwitch'in place names in the Great Bear Lake and Fort Norman regions. These semantic categories, he points out, may not have meaning to the Gwitch'in themselves, but serve as a research device for those outside the culture. They are included here simply to give non-Native readers a handle on understanding the scope of meanings embedded in the place names. The categories include:

- purely descriptive names,
- names associated with fauna or faunal activities,
- names associated with material culture,
- names associated with flora,
- names associated with historical events,
- names associated with mythological events,
- names associated with particular individuals,
- metaphorical names,
- names which reflect aesthetic qualities,
- names borrowed from other languages and
- names unanalyzable/untranslatable. (Andrews 1990: 8)

The place names provided in this discussion are a small sample to illustrate the literal definitions and the information conveyed in the names. These names have been taken from a number of sources. Some refer to the use of the land by foreign settlers, other name origins make it more difficult to determine what they refer to because of cultural change or change in use.

A number of names are incorrectly translated, such as Pne'katik (Benacadie, Cape Breton), translated by Rand as "the humble place" but meaning "the egg laying place." Other names most likely combine a number of attributes, both practical and mythological, as is often the case in the Mi'kmaw language.

Yet other names are adaptations of Basque or French words. Peter Bakker provides the example of the word *souriquois*, used by the early French explorers as a generic designation for the Mi'kmaq. The use of this term probably began on the isthmus between what is now Nova Scotia and New Brunswick, and an area where Samuel Champlain also mentions a "Souricoua River." Most likely the word comes from the Basque word *zurikoua*, which is glossed "white man's place," and may have indicated a place where Mi'kmaq would meet Europeans for trade (Bakker 1988: 7-15). Whitehead speculates that the local Mi'kmaq, speaking pidgin Basque to 17th-century French traders, used the Basque term *zurikuoa* to indicate the name of the river. The French then most likely expanded the word as a generic term for Mi'kmaq.

Place names in Mi'kmaw can be verbs or nouns. If the location were a place of a particular activity, such as eel spearing, the name would be a verb. If the area were a specific landmark, such as a lake or specific resource, it might be a noun. Two endings, which indicate that the place name is a noun, are *-a'ki or -e'kati*. The word "place" does not exist in Mi'kmaw, but is used as the closest English word to translate the meaning.

Wanpa'q: "calm water" (Cole Harbour, NS; verb)

Sipuk: "at the river" (Sydney, Cape Breton; noun) but used to be called *kun'tewiktuk* (noun) meaning "at the rock."

Wiaqajk: "the mixing place" (Margaree, NS; noun) The name connotes a place of ochre but *wiaq* suggests blending. The translation was provided by Wilfred Prosper, Eskasoni, with additional comment from Margaret Johnson and Bern Francis.

Tmaqnapskw: "something that looks like a pipe" (place in Miramichi, NB; noun) translated by Rand as "the pipe-stone place."

Tlaqatik: "at the encampment" (Tracadie, NS; noun).

Pankweno'pskuk: defined by Jerry Lonecloud as "lice-picking falls," but literally means "where they hunt one another's head [for lice]" (Gabriel Falls, NS; noun).

Penatkuk: "bird nesting place" (island in Shelburne River, NS; noun).

E'sue'katik: "the place of clams" (St. Esprit, Cape Breton; noun).

Sipekne'katik/Sikipne'katik: "where ground nuts are found" (Shubenacadie, NS; noun).

Amaqapskeket: "rushing over rocks" (Gold River, NS; verb).

Kukwesue'katik: "haunt of the giants" (Middle River, Sheet Harbour, NS; noun).

Kopitue'katik: According to Margaret Johnson and Wilfred Prosper, this means "place where there are lots of beaver or where beavers gather." (Beaver

Harbour, NS; noun). According to Silas Rand, "tradition has it that Kluskap threw one of the large rocks there at the mythical beaver."

Kespukwitk: defined by Wilfred Prosper as "the last flow of water" (one of the seven traditional Mi'kmaw districts, Southwest NS; verb).

Kjipuktuk: "the great harbour" (Halifax, NS; noun).

Epekwitk: defined by Wilfred Prosper as, "in the water, up above" (PEI and part of one of the seven traditional districts; verb).

Matuesuatp: "the head of a porcupine" (Porcupine Head, NS; noun).

Pne'katik: "egg laying place," translated by Rand as "the humble place" (Benacadie, NS; noun).

Waqmitkuk: "clean flowing water" (Wagmatcook, NS; noun).

Mntuapskuk: "Devil's Rock" (It was about 100 years from the time Maillard initiated a change in the meaning of the word Mnitu to mean "devil" instead of another word for "creator" (Jeddore, NS; noun; Rand/Francis).

Plekteaq: "split by a handspike" (This refers to the columnar rocks at Cape Split, NS; noun). Rand defined this as "a handspike" and recorded that tradition has it that Kluskap used one of these handspikes to open up the passage at Cape Split and drain the Annapolis Valley. Wilfred Prosper and Margaret Johnson said this "handspike" is something with a sharp end for moving heavy objects.

Napu'saqnuk (Rand: Naboosakunuk): "place for stringing beads" (St. Mary's, NB; noun).

Potlotek: May be derived from "Port Toulouse," the French fort erected in St. Peter's, Nova Scotia. (Chapel Island, NS; noun).

Tepotikkewey Qospem: originally translated by Margaret Johnson as "shaped like a boot." Linguist, John Hewson, stated that it is from the French "des bottes." He wrote, "to this has been added a locative -ek, so that Tepotek would be "boot place." Then when a common noun, "lake," is added, the original name is made into an adjective by adding -ewey. Presumably the full Mi'kmaw name is Tepotikkewey Qospem, Boot Lake, or rather "Boot Place Lake" (in Eskissoqnik, Eskasoni, NS; noun).

Tying in place-names and legendary sites show how the land is a visible, tangible part of the Mi'kmaw world view and cultural psyche. How this sentient landscape was formed is seen through legends and visible sites associated with acts of creation (e.g., Cape Split [Plekteaq], Partridge Island [Plawejue'katik] and many more). Place names tell of where to find resources, features of the landscapes, or where particular events took place. These legends, in turn, provide geographical information and, as will be illustrated in the following chapter, acted as oral maps of the landscape Mi'kmaq continue to call home.

CHAPTER 3
LEGENDS AS MIRRORS, MAPS AND METAPHORS

Storytelling was and is a traditional Mi'kmaw form of learning and communicating knowledge, as is true in many indigenous cultures throughout the world. As well, it was a source of power, a way to channel energy and shape reality. Information or messages could be, and still are by some, received through visions, dreams or experiences with the phenomenal world that included communication with animals, other-than-human beings such as *Wiklatmu'jk* (small people or fairies) or a host of other forms of energy. Telepathy and dreams are commonly mentioned in conversations even today. In traditional Mi'kmaw culture, these experiences were believed and trusted as potentially of importance for the information they carried.

There are two categories of Mi'kmaw stories. The first are termed *aknutmaqn*, meaning "news." The second are *a'tukwaqn*, meaning "stories treasured up," and indeed handed down from generation to generation, often told for diversion and to keep in memory the habits and manners, domestic and political, of the *sa'qewe'ji'jk* "the ancient ones" (Whitehead 1988: 221). As one Mi'kmaw told Wilson Wallis, regarding the origin of stories:

> Among the first generation of old-time Micmac there were no stories. The second generation told a true story about the first generation; the third generation made a story about the second, and added it to the other. The process continued and today a great many stories are known to us all. (Wallis and Wallis 1955: 317)

Perhaps "no stories" among the first generations shouldn't be taken too literally. Stories and story cycles have different functions. Many stories involve creation, animals, their behaviour and their interchangeability with humans; others are mapmaking and directional, giving geographical and resource information, as well as origins of natural features, as will be seen in this discussion. Some stories are moralistic in the sense of teaching the consequences of certain behaviours, or reinforcing values and/or told at a particular location where a mythical or historical event occurred.[1] Some stories may have been associated with different times of years, as will be seen in the example of the "serpent dance" in the discussion on dance, and the story of the Bear constellation later in this chapter.

Nicholas Denys wrote the following description of storytelling in the mid-1600s.

> The Indians were very fond of feats of agility, and of hearing stories. There were some old men who composed them, as one would tell children of the times of the fairies, of the Asses' skin, and the like. But they compose them about the Moose, the Foxes, and other animals, telling that they had seen some powerful enough to have taught others to work, like the Beavers, and had heard of others which could speak. They composed stories which were pleasing and spirited. When they told one of them, it was always as heard from their grandfather. These made it appear that they had knowledge of the Deluge, and of matters of the ancient Law. When they made their holiday feasts, after being well filled, there was always somebody who told one so long that it required all the day and evening with intervals for laughing. They were great laughers. If one was telling a story, all listened in deep silence: and if they began to laugh, the laugh became general. During such times they never failed to smoke…. Those story-tellers who seemed more clever than the others, even though their cleverness was nothing more than sportiveness, did not fail to make fun of those who took pleasure in listening to them. (Denys 1968: 418-19)

Storytelling is not just based on speech but is a multi-sensory experience, as will be further highlighted in the discussion on songs and dances in chapters 4 and 5. Similarly, information was received through all the senses, and was very important to survival in terms of tuning into to the environment, its smells, rhythms, sounds, textures and tastes. Tuning into and mirroring the sounds and rhythms of the world, and communicating through bodily movement, was an inherent part of Mi'kmaw culture and integral to storytelling. In 1758, Abbé Maillard documented a feast oration honouring the ancestry of the host:

> He was particularly admirable for decoying of bustards by his artificial imitations. We are all of us tolerably expert at counterfeiting the cry of those birds; but as to him, he surpassed us in certain inflexions of his voice, that made it impossible to distinguish his cry from that of the birds themselves. He had besides, a particular way of motion with his body, that at a distance might be taken for the clapping of their wings, insomuch that he has often deceived ourselves, and put up to confusion, as he started out of his hiding place. (Maillard 1758: 11)

In his journals from the late 1800s, Arthur Silver, wrote: "There is something peculiar to the Indian speech, abounding as it does in soft vowel sounds, which lends itself readily to the imitation of every sound of nature" (Parker 1995: 18).

The use of many sensory channels allows for information to be received and assimilated in a holistic and comprehensible way. This multi-sensory and extrasensory quality of experience in learning and communicating meant that the Mi'kmaq assimilated many layers of information, each of which informed them in important ways, and any one of which could mean life or death. For instance, smell was important to gathering of medicines and is still important for some Mi'kmaq, who smell plants and berries while walking in the woods. Some Mi'kmaw Elders smell people. Sweetgrass (*welima'qewey msiku*), burned as an offering and purification by Mi'kmaq, literally means "sweet smelling grass."

Silence also played an important role in tuning in and mirroring the many messages of the world (Sable 1996). Campbell Hardy's description of his Mi'kmaw guides during a hunting trip in Nova Scotia reinforces the role of silence and listening:

> A walk through the forest under heavy loades, is generally a tedious and silent affair. At long intervals, the Indians exchange a few syllables in their melodious language and impressive manner. Their subdued tones draw no echo from the woods, as does your quick and boisterous exclamation. Though they have no intention of hunting, should they find tracks quite fresh, their step is as light and their caution as unrelaxed as on the trail. In fact, either when hunting, or merely travelling through the woods, they avoid disturbing, in any way, game that might be in the neighbourhood of their route. (Qtd. in Parker 1995: 10)

Abbé Maillard remarked upon a seemingly more contemplative silence in an account of a feast in the mid-17th century:

> After grace being said by the oldest of the company, who also never fails of pronouncing it before the meal, the master of the treat appears as if buried in profound contemplation, without speaking a word, for a full quarter of an hour; after which, waking as it were out of a deep sleep, he orders in the Calumets, or Indian pipes, with tobacco. (Maillard 1758: 6)

Stories told by women most likely differed from those told by men. Women, although having integrated and overlapping roles with men (e.g., fishing and bird hunting), also had their own spheres of activity, which required different criteria for undertaking.

Dr. Margaret Johnson related a story about the creation of sweetgrass along the Bay of Fundy. She said she had originally heard this story from her father but that she had "made it better." One feature of the story, in her telling, was the description of a woman's hair, which was subsequently to become the sweet grass that grows along the Bay of Fundy shore. Johnson's description of the girl's hair talked of its beauty, its length and generally highlighted the hair as a central feature (Johnson, personal communication

1995). Hair, as seen in "The History of Usitebulajoo" (Wsitiplaju), was one of a woman's powerful aspects. It is the sister's hair greased with tallow that drew the puoin-bear's attention (Rand 1971: 44). First Nations traditionalists say that hair is very powerful and a man should only touch the hair of his wife and never the hair of other women.

Jerry Lonecloud, in the 1920s, remembered that the "Chief Medicine Man" would gather the young boys around him to tell the legends, but not the girls.

> Chief Medicine Man would once in a while (probably once a week) tell them about legends while the young Indian [*sic*] (young Indians only) sat around the wigwam. In those days they had a council wigwam where they held all their sayings and doings and it was in this council room these legions [legends] were told. Dr. Lonecloud [Dennis reporting] remembers this taking place 70 years ago in his childhood. They would be 70 or 100 young Indian boys present. Only Indian boys. The girls were not present. The women taught them. Dr. Lonecloud says they knew little of what the girls were taught but the girls were told the story of the canoe being 1st made by a woman—it was so smart in the women. (Dennis 1923: Notebook 1: 2)

Myths/legends are not fictitious; they are grounded in a reality experienced by each culture, an ongoing dialogue between a culture and the environment in which they dwell, an interpreting and reinterpreting of reality. As the anthropologist Irving Hallowell wrote:

> ...what people choose to talk about is always important for our understanding of them, and the narratives they choose to transmit from generation to generation and listen to over and over again can hardly be considered unimportant in a fully rounded study of their culture. When, in addition, we discover that all their narratives, or certain classes of them, may be viewed as true stories, their significance for actual behaviour becomes apparent. For people act on the basis of what they believe to be true, not on what they think is mere fiction. Thus one of the generic functions of the "true" story, in any human society, is to reinforce the existing system of beliefs about the nature of the universe, man and society. (Qtd. in Smith 1995: 19)

Mi'kmaw world view is an open system filled with possibility, mutability and ongoing interpretation based on personal and shared experience. Situations change and choices are made continuously, but cultural values and beliefs about reality are the mirror against which a person or culture makes these choices—they are the continuous reflection of who we wish to be and the reality we choose to create. As Theresa Smith wrote: "Myth ... is the inherently meaningful memory of the people spoken in the form of a symbolic narrative. It both defines and reflects reality and possibility in the world" (20).

LEGENDS AND STORIES
AS MIRRORS OF THE ENVIRONMENT

The following story, documented by Stansbury Hagar (italicized) with commentary by Bernard Hoffman (not italicized), has many levels of storytelling as well as the richness of information about the landscape embedded in the legends. Hagar's work, in conjunction with legends provided by Silas Rand and Jerry Lonecloud, illustrate the mirroring of the sky on earth, and earth in sky and how each reflected and embodied the other (Sable 1996).[2]

These stars and constellations are so arranged in the sky that the Bear:

...is represented by the four stars in the bowl of what we call the Dipper. Behind are seven hunters who are pursuing her. Close beside the second hunter is a little star. It is the pot which he is carrying, so that, when the bear is killed, he may cook the meat therein. Just above these hunters a group of smaller stars form a pocket-like figure—the den when the bear has issued....

The activities of these celestial characters were integrated by the Micmac in a legend which not only explained their relative positions in the sky, but also contained the motif of annual death and resurrection. In this case the celestial bear emerges from her den in the spring of each year, to be spotted and chased by the seven (the Micmac magic number) hunters. The chase goes on throughout the summer, and finally, in mid-autumn, the hunters who remain overtake their prey and kill her. Robin becomes covered with her blood in the process and attempts to shake it off, which he does except for a spot on his breast. The blood which he shakes off however

...spatters far and wide over the forests of earth below, and hence we see each autumn the blood-red tints on the foliage; it is reddest on the maple, because trees on earth follow the appearance of the trees in the sky, and the sky maple received most of the blood. The sky is just the same as the earth, only up above, and older....

After dancing around the fire and offering their thanks to the "Universal Spirit," the chickadee, the moose bird, and the robin feasted on their catch.

But this does not end the story of the bear.... Through the winter, the skeleton lies upon its back in the sky, but her life-spirit has entered another bear who also lies upon her back in the den, invisible and sleeping the winter sleep. When the spring comes around again this bear will again issue forth from the den to be again pursued by the hunters, to be again slain, but again to send to the den her life-spirit, to issue forth yet again, when the sun once more awakens the earth.... And so it is, the Micmacs say, that when a bear lies on her back within her den, she is invisible even to those who might enter that den. Only a hunter with great magic power could perceive her then....

[The Micmac] ... say that they know the Celestial Bear never dies, because she is always in sight, and that is why her earthly descendants never die of natural causes, but only fall asleep each autumn and come to life again in spring. For all earthly

animals are the descendants of the ancestor animal in the sky, and their appearance and habits are but reflection of hers. In all things as it was an is in the sky, so it is on earth.... (In Hoffman 1955: 252-54)

Robin (*jipjawej*), according to Lonecloud, is the very red star seen after 10:00 p.m., and is called such because of its red breast (Dennis 1925: Notebook 2: 99).

Lonecloud provides other information on the stars:

The seven stars are known by the Indians as the Bear's den. Hahjalquetch (possibly *qaqjalkwej*: Hewson, translation unknown) meaning seven stars. The bear having come to earth at the call of the young man to his pets. There are two stars that come out before daylight. (Ganet bird is this star ... Orion is three stars in a row supposed to be three chiefs fishing together. Each has a line consisting of a row of stars (three in a row, each has a line of stars from it). (Lonecloud in Dennis 1923: Notebook 2: 100)

Our forefathers were Stars. When they came upon earth, the woman star, she had bird pets which she dearly loved and she did not want to leave them up with the other stars, and she couldn't get back up there again. And she made a vow to remain on earth forever. She called her pets the birds. First the White Eagle, Fish Hawk, Ganet, [the list continues through a number of birds] ... bumble bee, yellow wasp, black hornet. These were stars in her days and she called them down. The man was longing for his animals and he called moose [*tia'm*] caribou [*qalipu*], bear [*muin*] ... (Lonecloud goes through numerous animals). (103-104)

There is also a Mi'kmaw song about the stars recorded originally by Mrs. Wallace Brown.

We are the stars who sing
We sing with our lights;
We are the birds of fire,
We fly over the night.
Our light is a voice;
We make a road for spirits,
For the spirits to pass over.
Among us are three hunters
Who chase bear;
There never was a time
When they were not hunting
We look down on the mountains.
This is the Song of the Stars.
(In Hoffman 1955: 349-50)

As a final note to this description, the Mi'kmaw word for Milky Way is *Skite'kmujuawti*, meaning "Spirit Road" (Hoffman) or "Ghost Path" (Hewson).

In the story of Muin and the bird hunters we see the earth mirrored in the sky, and vice-versa. Hagar himself wrote at length about the legend, noting how the story tracked the movement of the constellation through the seasons. He stated:

> When we attempt to interpret this legend, we cannot fail to be impressed by the singular fidelity with which its details present, often simultaneously, the habits of birds and animals and the movements of stars. Such accuracy, it is plain, can only result from long and careful observations of the objects described, and, indeed, whoever is acquainted with even our northern Indians knows well that very little in nature that can be seen with the naked eye escapes their observation. (Hagar 1900)

The interconnectedness between the movement of the stars, the changing of seasons, the hunt of the bear, the robin and the other birds, the trees and the celebration and honouring of all of this is embedded in this one story. Earth, stars, seasons, birds, stars, animals, trees and men move in synchronization—they are all related.

Specifically, the bear was the most honoured of all animals in Mi'kmaw culture, perhaps because of its human-like qualities, its power and as a provider of food and skins for shelter and clothing. The black bear (*Ursus americanus*) is omnivorous, feeding on vegetation and fish, insects and other meat. Like humans, they have a plantigrade, or a flat-footed walk, allowing them to place their foot flat on the ground and to stand up straight (Wernert 1982: 64; Wildlife Education n.d.: 3-4).

The black bear begins its preparations for hibernation in September, when it [he or she] begins to gain weight and collect leaves and tree branches, which it drags into its den. The den might be a hole in the ground, a cave, or under fallen trees or other protected areas. Nicholas Denys reports that the black bear hibernates in hollow trees (Denys 1968: 363). About a month later, as the snows begin to fall, the bears go into hibernation where it remains for up to six months. While in hibernation, they live off their stored fat. Denys reports that they suck their paws "for their entire living" while in hibernation (Denys 1968: 363; Wernert 1992: 64; Wildlife Education n.d.: 3-4).

 It is during the latter part of hibernation, usually in January, that the sows (female bears) give birth to their offspring. As Hagar's account of the story says: "Through the winter, the skeleton lies upon its back in the sky, but her life-spirit has entered another bear who also lies upon her back in the den, invisible, and sleeping the winter sleep" (qtd. in Hoffman 1955: 253). This probably is an allusion to the sow giving life to her embryonic cubs. The bear referred to in Hagar's rendition is a she-bear.

Denys relates that hunters might kill a bear in the winter if they happened upon it. ("Only a hunter with great magic power could perceive her then..."). Looking among large trees, they would search for "breath in the form of vapour" as an indicator for the presence of a bear (Denys 1968: 433-34).

Normally, bears would be hunted in the spring and through to the autumn by tracking them or with the use of deadfall traps. Denys relates another account, again involving breath, if the Mi'kmaw hunter is unsuccessful in taking the life of the bear.

> If the hunter does not bring it down, the Bear embraces him, and will very soon have torn him to pieces with his claws. But the Indian to escape this throws himself face down upon the ground. The bear smells him, and if the man does not stir, the Bear turns him over and places its nose upon his mouth to find if he is breathing. If it does not smell the breath, it places its bottom on the [man's] belly, crushes him as much as it can, and at the same time replaces its nose upon the mouth. If it does not then smell the breath, and the man does not move, it leaves him there and goes fifteen or so paces away. Then it sits down on its haunches and watches [to see] if the man does not move. If the man remains some time immovable, it goes away. But if it sees him move, it returns to the man, presses him once more upon the belly for a long time, then returns to smell at his mouth. (Ibid.)

In both cases, breath is used as an indicator of life and survival. Bear is mirroring man's behaviour and man mirrors bear's behaviour. This in turn is mirrored in the sky by the constellation of stars and in the story.

This story tells of the seasons, and the many indicators that tell that nature is shifting gears—the change in leaves on the trees, the migration of birds and their reappearance in spring (traditionally associated with robin red-breast), the hibernation of the bear and the movement of the Big Dipper throughout the winter. In the depths of winter, the bear is seen "on its back." Other information relates to the north star, the morning star, the red star (most likely Algol) and so forth.

Stories of people, animals and birds transforming from the stars are also not so far fetched when one realizes that the earth, as scientists have discovered, was formed from star dust. Similarly, as one realizes the earth is continuously in motion—tectonic plates are shifting, eruptions occurring, floods inundating vast tracts of land, soil eroding, novas exploding, life dying and decaying in a constant recycling of matter—one can see that the powers of tranformation are being wrestled with even today, only we don't relate to them as conscious beings.

LEGENDS AS MAPS

Much like place names, Mi'kmaw legends most likely served as "oral maps" of valuable resource areas as well as acted as accounts of natural events caused by climatic fluctuations resulting in changes to the landscape, such as the advancement of ice sheets and the opening of river valleys.[3] These legends contain a library of practical information encoded within mythological imagery and metaphors that provide knowledge of the environment and a narrative for how to "live right" as discussed in the previous chapters.

Examples of "oral maps" can be seen in numerous legends that refer to rock and mineral formations around the Minas Basin and Bay of Fundy. Piecing together the oral histories held by Mi'kmaw Elders, the legends recorded by missionaries, anthropologists and travel writers, archaeological and geological research conducted within Mi'kma'ki, and the linguistic analysis of place names, a map of lithic sources or lithic workshop areas (areas where raw materials were shaped into tools) emerges.[4] Sadly, a number of sites mentioned in the legends have not yet been excavated, and others have been inundated by rising sea levels over time.

SAMPLE LEGENDS

The following two legends refer to sites throughout the Minas Basin area. The first is taken from Silas Rand's *Legends of the Micmac* (1971), and the second from Jerry Lonecloud's accounts as transcribed in Clara Dennis's original field notes in the early 1920s. The first legend is entitled, "Wizard Carries off Glooscap's Housekeeper" and was told to him by a Mi'kmaw named Thomas Boonis of Cumberland County on June 10, 1870 (Rand 1971: 287). Only the portion of the legend concerning the Minas Basin area is included here, but other references to Cape Breton and Newfoundland are included in the earlier part of the legend. The place names are bolded for easy reference and emphasis, as well, the current spelling of the names in the Smith/Francis (S/F) orthography is included where possible.[5] There are various spellings of Kluskap (S/F) throughout different versions of the legends (e.g., Glooscap [Rand], Glooscup [Dennis] and Gluskap [Parsons]), which have been left in their original form, while using the Smith/Francis spelling, Kluskap, in general references to these legends. The Smith/Francis spelling has also been included with translations in parentheses following place names when possible, though some have not yet been glossed. Who Kluskap is, or was in the pre-contact period, is another area of research interest for Mi'kmaw and non-Mi'kmaw scholars alike.[6] We do know he (she/it?) was a powerful, creative force associated with many landscape formations, as well as other resource areas.

> ...The next adventure mentioned in our narrative occurred at Partridge Island [Plawejue'katik: "partridge place"]. Here he met with another worthy, of unnatural birth and supernatural nurture, and of vast supernatural powers. His mother fell a prey to the cannibal propensities of an ugly giant; and he was taken alive from his mother after her death, thrown into a deep spring, where alone and unattended he came to maturity, and afterwards came forth from his place of concealment to avenge the death of his parent, and to go forth as

a deliverer of the oppressed and a general benefactor to his race. His name, which describes the manner of his birth, was Kĭtpooseăgĭnow.

Glooscap halted at the lodge of this personage (it were hardly fair to call him a man), and he proposed to his guest in the evening to go out fishing by torchlight. The canoe, the paddle, and the spear were all made of stone. The canoe was large and heavy; but Kĭtpooseăgĭnow tossed it upon his head and shoulders as though it were made of bark, and launched it into the bay. As they stepped on board, Glooscap asked which should take the stern paddle, and which the prow and the spear. (They catch a whale, go back and roast it.)

Before going farther up the bay, [Kluskap] now crossed over to Utkogŭncheech [Tkoqnji'j: Cape Blomidon: also Metoqwatkek: "bushes extending down the bank"]. There he arrayed his aged female companion, decked her with beautiful beads and strings of wampum, making her young, active and beautiful, and for her sake making all those beautiful minerals for which the "hoary cape" has been so long celebrated. My aged friend Thomas Boonis, who related this narrative to me, assured me with much animation that he had seen these beautiful minerals with his own eyes,—emphasizing his assertion by saying in broken English, "Glooscap, he makum all dese pretty stone." I allowed the worthy man to enjoy his own opinions without a lot of hindrance from me, only urging him to hasten on to the end of his tale....

His next halt was on the north side of the bay, Spenser's Island [Wtuoml: "his/her pot"]. There Glooscap engaged in a hunting expedition on a somewhat large scale. A large drove of animals was surrounded and driven down to the shore, slaughtered, and their flesh sliced up and dried. All the bones were afterwards chopped up fine, placed in a large stone kettle, and boiled so as to extract the marrow which was carefully stored away for future use. Having finished the boiling process, and having no further use for the kettle, he turned it bottom upwards and left it there, where it remains in the form of a small round island, called still by the Indians after its ancient name, Ooteomŭl ("his kettle": that is Glooscap's kettle) [Wtuoml: "his/her pot"].

He now visited a place lying between Partridge Island and the shores of Cumberland Bay, and running parallel to the River Hebert. It is called by the Indians Owkŭn [Wo'qn: "spine"] but in English, River Hebert.

He now pitched his tent near Cape d'Or [L'mu'juiktuk: "place of the dogs"], and remained there all winter; and that place still bears the name of Wigwam (House). To facilitate the passing of his people back and forth from Partridge Island to the shore of Cumberland Bay, he had thrown up a causeway, which still remains and is called by the white people "the Boar's Back." It is this ridge which gives the Indians name Owŏkun [Wo'qn] (means crossing over place) to the place and to the river.[7]

In the ensuing spring, while he was out hunting with his dogs, a moose was started, and his dogs pursued him to the land's end at Cape Chignecto [Sikniktuk: "the draining place"]. There the moose took to the water and struck boldly out to sea, whither the dogs with all their magic, could not pursue him. But they seated themselves on their haunches, raised their fore-paws, pricked forward their ears, and howled loudly and piteously at the loss of their prey. Glooscap arrived on the spot in time to witness the interesting spectacle. He stopped the moose and turned him into an island, which is known as the Isle of Haut [Maskusetkik: translation uncertain]; changing the dogs into rocks, he left them there fixed in the same attitude, where they are to be seen this day, watching the moose.

Near Cape d'Or [L'mu'juiktuk] he fed his dogs with the lights of the moose; large portions of this food were turned into rocks and remain there to this day; the place is called Oopunk (Wpnk: "his or her lungs"). Glooscap now took the old woman and set her down and, telling her to remain there, he turned her into a mountain, which is to be seen to this day; but he told her that when he reached his island home in the far west, she would be there with him. He then left the country, and never came back to it again. He went on to his beautiful isle in the west; and when he arrived and had fixed his dwelling and furnished it, there in her place was found his faithful housekeeper and her little attendant, Marten. (Rand 1971: 290-93)

From Clarissa Archibald Dennis's manuscripts documenting Jerry Lonecloud:

Glooscup's camp was the point at Advocate Harbour [Atuomjek: "at the sandy place"]. His stationary wigwam was at Advocate Harbour. That is where he lived. There was a Beaver Dam there from Cape Blomidon to Spencer's Island [Wtuoml: "his or her pot"] (about 9 miles across). Glooskup being a great hunter wanted to kill one of these beavers for his food. He set a deadfall trap (a wooden trap) on Blomidon. He did not catch any; they wouldn't go in the trap so he took his bow and stuck it above the upper part of the beaver dam. With that he slewed the Beaver Dam out of business. So he thought his plan was to kill one of the Beavers as the water was leaving Truro Bay or Basin [We'kopekwitk: "end of the flow"], whatever it is called. The Beavers riled the waters up in Truro Basin—made it muddy looking so the Beavers escaped unseen by him. (This is why Truro Bay is so muddy. The Beavers weren't out in the Bay of Fundy.)

Then Glooscup went to Saint John to the falls there. He thought the Beavers would go up Saint John's River [*sic*] and he would head them off. They didn't go up there. He looked at the Briar Island—it was a point at that time. He saw a Beaver going over the neck of land. He picked up a stone, fired at the Beaver and missed it (forty miles away it was) and made a channel. The stone cut the neck of land and made it a little passage called Petite Passage [Tewitk: "flows out"] now. The other Beavers—there were two (he missed the first) went to the western part of the Island and at Briar Island Passage, made a passage and is now that island, and that Island is the stone thrown from Grand Falls [Kji-qapskuk] (it is a little Island with a light house on it.) He missed both Beavers and returned to his camp at Advocate and there was some beautiful stones there—amethysts. These he made beads of and wore them for good luck.

Then he went to Cape Breton Island [Unama'kik: variation of *Mi'kma'kik*: "place of the Mi'kmaq"]—leaped over. Here he discovered the same Beavers. He killed the smallest Beaver with bow and arrow and had a great feast by himself (he always lived alone). Then he was satisfied with what he had got for all his trouble. (A bone found here of the animals existing here before the flood is in the museum and is supposed to be one of the bones of these beavers.) [This is the mastodon's thighbone according to Ruth Whitehead]. Then he went back to Advocate to his old camp again. Where he got his water to drink was at Parrsboro [Taqamiku'jk: "little causeway" or "little crossing place") fifteen miles from camp, from a lake called Glooscup Lake. When he came back he told the people and told his dogs (he had two dogs) he was going away north and "I will come back at the end of the world. I am going to make you a happy hunting ground." He says to his dogs, "Now we will have a moose chase." They chased the moose—they calculated to kill the moose but didn't and the dogs chased him in the water of Advocate Harbour and the moose was swimming out toward Isle of Haut or Spencer's Island. And when Glooscup came to the shore, he say to the moose, "I am going to leave you here for a landmark. You turn to stone, Moose. And there was, until twenty years ago, a stone island a perfect shape of a moose but 20 years ago the head of the moose disappeared owing to storms etc.

Glooscup went back to camp without any moose and he went up to get some water out of Glooscup Lake and he saw a partridge when he was getting the water. He didn't have his bow and arrow and he took a stick and chased the partridge on to the shore of Truro Basin or Cobequid Bay [We'kopekwitk: "end of flow"]. The partridge waded out into the water and Glooscup couldn't reach it to hit it with the stick he had. "Now," he says to the partridge, "now I'm going to leave you for a landmark. You will be an Island and your feathers will turn to trees." They call it Partridge Island in Parrsboro. Then Glooscup went back to his wigwam at Advocate Harbour. He told his people and his dogs, "Now I am going away to leave you." Then he went North. [Dennis 1923: Notebook 1: 135-238]

Further references to place names around the Minas Basin can be found in Elsie Clews Parson's research conducted in the 1920s, in "Gluskap's Moose Hunt" (Parson 1925: 86).[8]

GEOLOGICAL INFORMATION

In Lonecloud's story, all the places named, with the exception of Saint John and Cape Breton, are of the same ancient rock formation stretching from Briar

Island to the hook at Cape Split [Plekteaq: "split by a handspike"], where it then goes under water and across the Minas Basin to re-surface at Cape d'Or. Additional outcrops of the formation occur at Partridge Island, Five Islands, which includes Moose Island and Bass River [Ji'ka'we'katik: "at bass river"] (Goldthwait 1924: 18, 22).[9] This ancient rock formation, known as North Mountain, is rich in high-quality, lithic source materials—chalcedonies, agates and jaspers—used traditionally by the Mi'kmaq for making stone blades and tools. Chalcedonies are only found in this area of Nova Scotia and the whole Parrsboro area is known today as a "rockhound's" paradise because of these and other minerals, including amethyst, found there.

The Mi'kmaq preferred these materials because their fine, crypto-crystalline structure could easily be "flaked" with the tip of an antler and stone hammer to make sharp edges for cutting, scraping and penetrating hides. These minerals are quite beautiful, with swirls of colour in chalcedony, or deep, waxy reddish colour as seen in jasper (Whitehead, personal communication 1993).

Both the historical and archaeological record support the use of a number of areas by the Mi'kmaq around Minas Basin up to the late, prehistoric period as workshop or quarry sites for these minerals. These sites include Partridge Island, Five Islands (including Moose Island) Spenser's Island and Scot's Bay on the Blomidon Peninsula. Jasper is also found at Isle Haute, and Cape d'Or is also known as a source of native copper (Booth n.d.: 78, 84).

Archaeologist Micheal Deal excavated a quarry site dated to 1540 BP in the Scots Bay area (Scotch Bay in Parson's recording of the legend "Gluscap's Moose Hunt," previously cited) along the Blomidon peninsula. At this specific quarry site, rough "blanks" were manufactured by Mi'kmaq from the chalcedonies found in outcrops along the shore. These "blanks" would be brought back to encampments or distributed in trade, and later refined into tools and blades.

Deal's research into this quarry site, supplemented by archaeological and geological research into the whole Minas Basin area, has led him to hypothesize that two distinct lithic distribution routes existed throughout this area (Deal 1989: 3-5) (see figure 11).[10]

Because of the large number of unfinished "blanks," Deal goes on to explore the possibility of this being a quarry site for an intraregional lithic exchange network controlled by the local band, versus a regional quarry site. He speculates that in the pre-contact era, summer excursions would be made to the sites on the Fundy shore to manufacture quarry blanks and finished tools. These would then be taken back to summer camping sites, such as Melanson on the Gaspereaux River and St. Croix, where they could be distributed or exchanged with other local bands within the district. Summer camping sites were an ideal place for exchanges to take place; summer

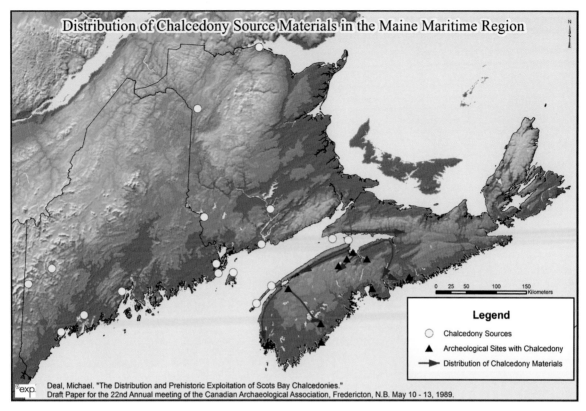

Figure 11
Dots indicate "general distribution of chalcedony source area throughout the Maine-Maritime Region. The black triangles on the map indicate archaeological sites in western Nova Scotia where chalcedony from Fundy shore sources have been identified. The arrows indicate the most likely distribution routes for Fundy shore chalcedonies" (Deal 1989: 4). Map compiled by William Jones and Trudy Sable. Data adapted from content contributed by M. Deal. Sourced by exp. Architects Ltd., NSGC and Geobase. Basemap layer ©2009 ESRI and USGS.

was traditionally a time of year when larger encampments of Mi'kmaq were formed for purposes of socializing, marriages and council meetings regarding peace and war (Deal 1989: 4-5; personal communication 1994).

Deal proposes that the lithic evidence found at this workshop site "can be viewed as the initial stages of a tightly integrated system that involves the selection, modification, distribution and consumption of lithic materials" (Deal 1989: 2). Archaeologist David Keenlyside has speculated that a large network of material exchange was present in the southern Maritimes from approximately 700-1200 AD. These trade patterns most likely continued up to the period of European contact. Many sites, however, have been lost to extensive erosion along the shore (Deal, personal communication 1994).

When the sites mentioned in the legends are put together with the archaeological mapping of quarry sites (see figure 12), it seems highly probable the legends recounted and informed people of excellent quarry

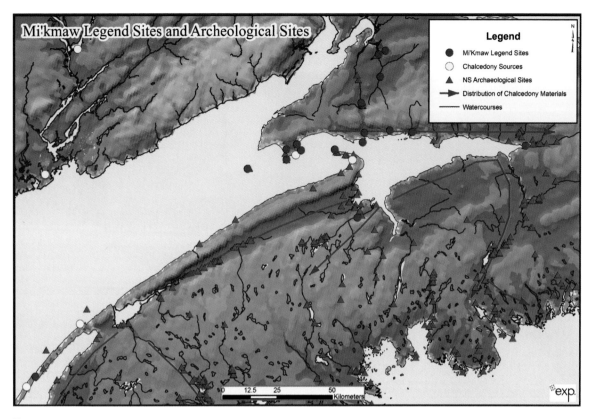

Figure 12
Map compiled by William Jones with data adapted by content contributed by Michael Deal, Roger Lewis and Trudy Sable. Sourced by exp Services Inc. Base map layers c 2012 ESRI.

sites—and possibly delineated distribution routes—and/or may have been territorial district markers. In short, they acted as maps.

THE BOAR'S BACK

The Boar's Back was also referred to in Rand's recording of the legend, "Wizard Carries off Glooscap's Housekeeper." As noted earlier, Rand interpreted Owōkun as "the carrying over place." As seen from the previously cited legends, this is Wo'qn and translates as "spine." Father Pacifique, the Capuchin missionary among the Mi'kmaq in the 1890s, glossed the Mi'kmaw name for Parrsboro as Ogotagamigotjg meaning "causeway." In the Smith/Francis orthography, it is taqamiku'jk, which is translated "little causeway" or "little crossing place."

A "boar's back" is a layman's term used for a rock formation such as this stone ridge that acts as a "causeway." Viewing a satellite image of the area created by geomatics expert William Jones, this landscape feature is also at the confluence of the two rivers—River Hebert and Maccan River. Visually,

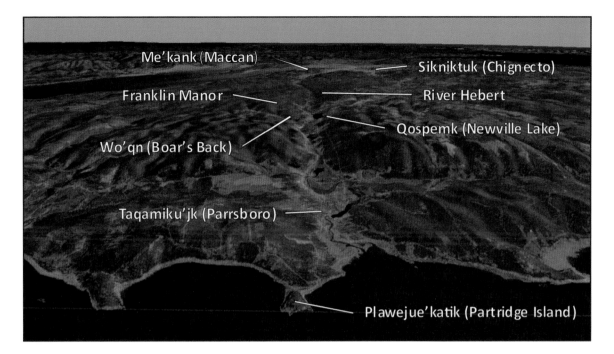

Me'kank (Maccan)   Sikniktuk (Chignecto)

Franklin Manor   River Hebert

Qospemk (Newville Lake)

Wo'qn (Boar's Back)

Taqamiku'jk (Parrsboro)

Plawejue'katik (Partridge Island)

wo'qn or "spine" could refer to the rock formation looking like a spine, or as the whole river system that runs from Parrsboro to Cape Chignecto and acted as a lifeline so to speak.

Elder Doug Knockwood, of Indian Brook (Sipekne'katik) First Nation, grew up on Lake Newville. His grandfather Sam Knockwood owned land there during his lifetime and there was a reserve named Franklin Manor established in 1865.[11] Mr. Knockwood's uncle was one of the last people to live on the reserve.[12]

> Franklin Manor is the reserve in Cumberland County connected to Newville Lake. My grandfather, I don't know why, bought property in around this lake. He owned property on that lake. I don't know how many families lived on it but there were a few. This man lived in Franklin Manor almost all his life and he never spoke a word of English, but he survived and made a living. (Knockwood, interview 2005)

Mr. Knockwood remembered reading a story about a French and English war on the boundary between Nova Scotia and New Brunswick. He had also heard a story that during this war, Mi'kmaq were popping up everywhere in the area.

> They didn't know how the Mi'kmaq were getting there. They told me there is a tunnel, caves from the Minas Basin to the Cobequid Hills and this is how the Mi'kmaw people travelled from the shores of Parrsboro to the marshlands of New Brunswick. (Knockwood, personal communication 2005)

Figure 13
Map compiled by William Jones with data adapted by content contributed Trudy Sable and Bernie Francis. Sourced by exp Services Inc. Base map layers c 2012 Esri.

Figure 14
Sam Knockwood,
Grandfather of Doug
Knockwood by Lake
Newville. Permission of
Doug Knockwood.

Figure 15
Doug Knockwood. Photo
by Trudy Sable.

Given the longevity of Mi'kmaw presence in this area, there is undoubtedly a rich oral and archaeological history that could be unveiled with further research.

BLOMIDON AND THE "EYE OF KLUSKAP"

The legendarily, "eye of Kluskap" (Kluskap *wetapit*: "from where Kluskap sees") is found at Cape Blomidon. Jim Simon of Shubenacadie knows of it, but little else is recorded. Ian Booth speculated that it could be a wind hole a couple of meters wide formed by currents eroding the cliff through Little Cove at Cape Split through to Amethyst Cove. The wind hole has the appearance of an eye of a needle. Alternatively, there is a beautiful mineral called faroelite, a "rare phenomenon" stone with shifting chatoyant (cat-like, ed.) eyes that has a "truly uncanny appearance when found loose on the Fundy shore as highly polished beads" (Booth n.d.: 30). Faerolite is found very near this wind hole. Both agate and faroelite can be found at the nearby Black Hole Fiord located on the south side of Scots Bay (ibid.).

This area is also the location of a beautiful amethyst formation at Amethyst Cove. As mentioned in the legend recorded by Rand in the late 1800s, "Gluscap goes from Partridge Island to Cape Blomidon, where he "decks his aged female companion out in beautiful beads" (Rand 1971: 291).

As mentioned in the previous chapter (see p. 42), Cape Blomidon is a "Grandmother" site, so perhaps the reference to his "aged female companion" concerns this distinctive rock face itself embedded with beautiful minerals. Perhaps these minerals were the amethysts, or other minerals found in the area, such as faroelyte, which can be found near Cape Split. Similar information can be illustrated relating to Isle Haute,[13] Cape d'Or, [14]Advocate Harbour (Kluskap's medicine garden), and other locations throughout Mi'kma'ki.

CAPE BRETON

In Cape Breton, Parson's recording of "Gluskap and Beaver" includes Salt Mountain (Wi'sikk), Indian Island, Little Island and Grand Narrows.

> From Salt Mountain [Wi'sikk: "shaped like a beaver's den"] [Parsons 1925: n. 4: "The English name is from the fact that the water out of the mountain is salty. The meaning of wi'sik was unknown to Mrs. Morris. To her it did not mean cabin" (Speck 1: 59)] Gluskap was chasing a beaver; beaver made holes in Indian Island (Elnuwe'e minigu) [L'nui-mniku] trying to get under it.... He did get under, went to Elguanik. Came out at Tewil [Tewitk] (Grand Narrows). The rock Gluskap threw at Beaver became Little Island [Mniku'ji'j]. The lesser of the two elevations of Indian Island was also made by the soil thrown by Gluskap.(Parsons 1925: n. 6 "Rand 1: 216"). From Salt Mountain Gluskap could make Indian Island in one step (i.e., stand with one foot on the mountain, the other on the island).

Little Island is said to consist of a rock rising about ten feet out of the water. But as it is still under the spell of Gluskap (My term, Morris's used various paraphrases, e.g., "whatever Gluskap says is true.") if anybody goes up on that rock, "trying out Gluskap's word," he feels as if he were held fast. Mrs. Morris's grandfather, Peter Newell, climbed this rock and there he was, "couldn't come down. Old lady had to push his eel spear up to him, got him down." He reported that the world had seemed very distant, far below him, out of reach. (Parsons 1925: 86)

The name Parson's gave for Salt Mountain was Wi'sikk, but the translation of the name was uncertain until a recent interview with Gregory Johnson of Eskasoni. Johnson said the *wi's* part of the word referred to beaver den, while the "*ikk*" part of the word meant it was shaped like a beaver den, so the name, now transliterated as Wi'sikk (S/F), would mean "shaped like a beaver house or den."

In a subsequent interview with Florence Young of Eskasoni, she said that when she was growing up, she was told that Salt Mountain was a sacred mountain and that you never should insult or speak badly of it because it would bring bad weather. She said that people could tell the weather by looking at the mountain, such as if the top of it was clear or shrouded in mist.

Weather mountains exist among the Innu of Labrador and are still respected today. During a recent trip to Kamestastin in Northeast Labrador, people were continually warned not to point at such a mountain because it would bring bad weather.[15] A number of these mountains are known by the Innu throughout Labrador, and most likely existed for the Mi'kmaw as well, though few have been identified to date.

Figure 16
Gregory Johnson of Eskissoqnik (Eskasoni) First Nation.

GEOLOGICAL FORMATIONS AND GLACIAL ACTIVITY: THE ANNAPOLIS VALLEY AND GASPEREAUX

There are a number of interesting references to landscape changes, such as the opening up of the Minas Basin and the draining of the Annapolis Valley. For instance, Cape Split is another landmark where Kluskap built deadfall to capture the beavers. Rand documented the place name, Plekteok (Plekteaq: "split by a handspike"), which he glossed as "huge handspikes for breaking open a beaver-dam" (Rand 1875: 85). Legend has it that Kluskap used one of these handspikes to open up the passage at Cape Split and drain the Annapolis Valley. These columnar rocks can still be seen and are distinct landscape features as seen from across the Minas Basin.

The 19th-century missionary, Father Pacifique, wrote:

Klooscap, the Indian demi-god, lived at Blomidon in a very large wigwam; Minas Basin was his beaver pond, for he had everything on a large scale; the dam was at Cape Split. But the Giant Beaver having become over proud, he cut it open. (Pacifique 1935: 289)

The Minas Channel was formed millions of years ago during the Triassic/Jurassic. However, during the Younger Dryas period, approximately 10,800-10,400 BP (approximately 12,900-11,000 calibrated) a period of substantial cooling occurred causing the PaleoIndians occupying the Debert area to retreat from the region. Again ice sheets formed, and the world went into a minor glaciation. The sea levels may have fallen, and then risen again following the retreat of the ice sheets. As ice sheets melted, numerous lakes were formed in the region (Fader, personal communication 1996).

A legend told by Stephen Hood in 1869 and transcribed by Rand entitled, "Glooscap, Kiwkw and Coolpujot" (Rand 1971: 236-37)—the following segment just part of a much longer story—about Mi'kmaq seeking out Kluskap to have their wishes fulfilled. It involves other place names such as the Bras d'Or lake [Pitu'pok] in Cape Breton [Unama'ki], Cape Chignecto [Sikniktuk: "the draining place"], Isle of Haute [Maskusetkik], Ooteel [Wti'l] (a plain near Isle Haute), and Spenser's Island [Wtuoml].

> In former days, water covered the whole Annapolis Valley [Tewopskik: "flowing out"] and Cornwallis valley. Glooscap cut out a passage at Cape Split and at Annapolis Gut, and thus drained off the pond and left the bottom dry; long after this the valley became dry land. Aylesford Bog was a vast lake; in this lake there was a beaver-house; and hence the Indian name to this day—Cobeetĕk ("at the Beaver's home") [Kopitek: "place of the beaver"]. Out of this beaver's house, Glooscap drove a small beaver, and chased it down to the Bras d'Or [Pitu'pok: "long dish of salt water"] lake in Cape Breton [Unama'kik]—pursuing it in a canoe all the way. There it ran into another beaver house, but was killed; and the house was turned into a high peaked island; Glooscap feasted the Indians there. A few years ago a heavy freshet tore up the earth in those regions, and laid bare the huge bones of the beaver upon whose flesh Glooscap and his guests had feasted,—monstrous thigh-bones, the joints being as big as a man's head, and teeth huge in proportion. (Rand 1971: 236-37)

Stea's research regarding the re-advancement of ice sheets during the Younger Dryas period in northern Nova Scotia and the Gulf of Saint Lawrence, and the subsequent damming of the Shubenacadie River area, offers a tantalizingly possible correlation to the legend. His work shows that a glacial lake filled the Shubenacadie valley, and the lowlands that surround it, up to 30 metres deep.

> It was formed when advancing and retreating glaciers dammed the present outlet of the Shubenacadie River in the Minas Basin. Rather than draining northward, the glacial lake drained southward through the route of [what we know as] the Shubenacadie canal system into Halifax Harbour. A circular pond that marks the upper reaches of the sand body ... is probably a sinkhole.

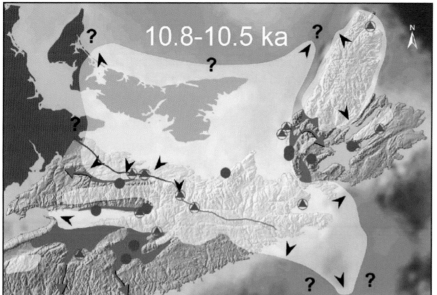

Figure 17
Reconstruction of the re-glaciation of the Minas Basin area from the work of Ralph Stea. The white shows ice cover, and the blue indicates glacial lakes. The black arrows indicate the direction of melt water through spillways (channels), the red dots are where glacial lake sediments were found overlaying peat, and the red triangles represent sites where glacial till was found overlaying peat. (Stea and and Mott 2005: 345-62).

The most consistent interpretation is that this sand body developed along the shoreline of a glacial lake (Glacial Lake Shubenacadie 2) fed by meltwater streams from a glacier to the east. The lake re-formed in the Shubenacadie Valley after drainage of the former, larger glacial lake (Glacial Lake Shubenacadie 1). These ice-dammed lakes formed when normal drainage via the Shubenacadie River toward the Bay of Fundy was blocked by residual or re-advancing glaciers during the Younger Dryas climatic event. (Nova Scotia 2011)

Stea also described how outlets became open with the retreating ice. Because the ice is retreating toward the Truro Basin, the first thing that comes open is the Minas Channel when the ice retreats far enough. The melt water then drains out, which causes the ponding up of lake water in the Shubenacadie Valley. However, there still may be ice in the eastern Minas Basin, which drains when the Shubenacadie outlet comes open (Stea, personal communication 2009; Sable 2011: 167).

In the late 1600s, while residing in the Gaspé Peninsula of Quebec, Father Chrestien Le Clercq documented a Noah's Ark-like story of a flood, which covered the whole earth. Many Mi'kmaq perished in the raging waters and high winds while trying to escape in their bark canoes. Only a kind and virtuous old man and woman survived to live long and fruitful lives and re-inhabit the earth (Le Clercq 1968: 84-85). Could legends, such as this one recorded by Le Clercq, be regarded as first-hand accounts of events that took place?

Debert itself is situated at the head of the Cobequid Bay; its habitation is dated to 10,500 BP. Artifacts from the site show lithic material most likely coming from Scots Bay. During this time, the continental shelf was exposed, and the landscape went through dramatic changes with the retreating and re-advancement of glaciers.

With careful research under the guidance of Mi'kmaw elders, along with the re-reading of the numerous legends and documentation of oral histories, we may be able to understand how these legends reflected and documented dramatic weather changes and subsequent landscape transformations.

IMAGERY: BEAVERS, OVERTURNED POTS AND WIKUOML (WIGWAMS)

The power of these legends comes through understanding how they communicate multiple layers of information encoded in the language, the imagery and the use of metaphors. Beavers have been used throughout these legends to talk about changes to the landscape. The dam of the beaver is burst open at the area stretching across the bay from Cape Blomidon to Spenser's Island; the tail of the beaver is in the Cobequid Basin where the beavers riled up the waters and made them muddy; Salt Mountain is shaped like a beaver den.

Ruth Whitehead, curator emeritus to the Nova Scotia Museum, has conducted research in the study of the giant beaver, *Casteroides ohioenses* that once existed in Nova Scotia. These animals, cousins to the modern beaver, were two metres tall and three metres long (eight feet tall and ten feet long); they died out after the last ice age, approximately 10,000 BP. Beavers are featured in a number of legends describing the creation of the landscape and waterways throughout the Maritimes. The giant beavers may have been models for such stories (Whitehead, personal communication 1995).

Whitehead also points out that it is in the nature of beavers to change the landscape. They fell trees and mound earth, sticks and trees to make their dams. Beaver teeth were also used for carving and incising.[16] This may be why Giant Beavers are found in many creation myths throughout Algonquian legends regarding the formation of the landscape. The Innu of Labrador, for instance, have similar legends.

Along with beaver metaphors, Kluskap legends from different regions also have similar story lines and imagery.[17] For instance, Bonaby Rocks in New Brunswick are also associated with Kluskap's overturned kettle, similar to Spencer's Island in Nova Scotia. More fascinating is that this is a fossil area that looks like the lower backbone of a fish but is actually "fossilized plant stamps from that particular type of coral that grew here millions of years ago" when the land was submerged (Allen 1997). From her discussions with Margaret Labillois, Elder from Eel River Bar First Nation, Archaeologist Pat Allen learned that the Mi'kmaq interpreted these fossils as fish bones, and associated this rock formation and the fossil beds with a legend of Kluskap turning over his pot in annoyance with the missionaries changing his peoples' ways. The fish bones are from the fish Kluskap ate during his last supper before overturning the kettle (Allen 1997, personal communication 2006).

From Rand we see the story associated with Spencer's Island.

> His next halt was on the north side of the bay, Spenser's Island. There Glooscap engaged in a hunting expedition on a somewhat large scale. A large drove of animals was surrounded and driven down to the shore, slaughtered, and their flesh sliced up and dried. All the bones were afterwards chopped up fine, placed in a large stone kettle, and boiled so as to extract the marrow which was carefully stored away for future use. Having finished the boiling process, and having no further use for the kettle, he turned it bottom upwards and left it there, where it remains in the form of a small round island, called still by the Indians after its ancient name, Ooteomŭl (his kettle; that is Glooscap's kettle [Wtuoml: "his/her pot"]. (Rand 1971: 292)

Allen's work presents another interesting possibility, which is that these sites are associated with fossil beds such as occur in the Parrsboro area, or that fossil beds or other distinctive features, are integral to the narrative and present visual imagery that can easily identify a place.[18]

Another legend documented by Allen is associated with the "flower pots" in Hopewell Rocks Provincial Park in Shepody Bay, New Brunswick. Allen was struck by these prominent, unusual and naturally occurring rocks, and sought out Michael Francis, an Elder and storyteller from the Big Cove First Nation [L'sipuktuk] in New Brunswick, for the oral history associated with it. Francis explained that the flower pots are just the name used in the tourist brochures, but were really cooking pots used by Ginaps [Kinapaq], or powerful Mi'kmaq, during an annual gathering of Mi'kmaq who came from all over the Atlantic provinces.

> For hundreds, perhaps thousands of years, before the arrival of the Europeans, the very powerful, very wise men of the Mi'kmaq people would gather together annually during the fall natural harvest. These men were called Ginaps. They would travel to the place of their "cooking pots" guided to this location by six foot high carved poles (wass geige) [l'nuoqta'wk: "carved poles"]. Mi'kmaw men, women, and children traveled long distances to come together for feasting, dancing, singing, and spiritual ceremonies. Food for all the people was provided by the Ginap who would prepare everything that was needed in their large "cooking pots."

> This annual gathering was gathered on for centuries until the European missionaries arrived. They convinced the people to take down the carved guiding poles saying that their enemies from the west would surely find them if the poles remained standing. With the pulling out of the sign poles and with the dying off of the Ginaps, the gatherings at the cooking pots ceased.

The big pots themselves turned to stone. We can still see them today as the Hopewell Rocks or *flowerpots*. The Rocks have remained in Mi'kmaw memory as a special place to go to meditate and to pray, especially if there was a food shortage. (Francis as qtd. in Allen, Nicholas and Thériault 2004: 9)

This oral history further supports the power of legends and their connectedness to the landscape, in this case, to rock formations. The "flower pots" are sandstone pillars eroded away by tidal action. The tides rise to 12.80 metres, and there are only six hours during the day before and after low tide when people can safely access the beach.

From Michael Francis's account, this was a gathering place and a place of power. It was a place where people shared information and were nurtured spiritually and physically. In the imagery of the overturned pots, we hear of the ending of a cultural period brought by a new religion, Catholicism, which marked a time when Kluskap and other powerful Mi'kmaw Kinapaq seemingly went away.

Figure 18
John Nicholas Jeddore
with eel spear.

LANDSCAPE AS WIKUOM

Looking deep into the legends, the landscape is metaphorically and literally perceived as home, or a large *wikuom*, filled with cooking pots (some overturned), relatives, leftover foods, tools and medicines. This perception of the landscape as "home" is at the heart of *weji-sqalia'timk*.[19] There are still Mi'kmaq today who live this belief. They include John Nicholas Jeddore (figure 18) the late Nikkli Jeddore and the late John McKewan from Miawpukek First Nation (Conne River) Newfoundland, and John McEwan from Maupeltu (Membertou) First Na-tion in Cape Breton.

These men were not threatened by the forest or the animals or the weather and feel at home everywhere on the land. They could bed down for the night wherever they found themselves even if they were in the forest after the sun went down. "How can anyone ever be lost in the woods?" as stated by the late Matthew (Matty) Jeddore.

CONCLUSION

Legends and oral histories correspond with current archaeological and geological knowledge regarding the landscape throughout Mi'kma'ki. Michael Deal's mapping of lithic sources around the Minas Basin augmented by the excavation of specific sites such as Isle Haute, Cape d'Or, Scots Bay, Melanson, St. Croix, etc., reveals the depth and extent of land use and resource networks throughout the Minas Basin area. Much more can be done to connect the information with legends from other areas, such as the travel and trade routes from New Brunswick into present day Nova Scotia or the Boar's Back as part of a travel route to Chignecto. Focusing on any area throughout Mi'kma'ki undoubtedly would yield extensive information.

Whether oral histories and legends can be accurately pinpointed to actual geological events requires far more research, but Ralph Stea's work regarding reglaciation during the Younger Dryas again points to the potential. What we do know is that encoded within the legends lies a storehouse of information and insight regarding the land formations and natural processes that formed the landscape of what is now called the Atlantic Provinces and into Quebec, and that these legends most likely also served as maps. As a whole, these legends show enough similarities in themes and imagery, such as Kluskap overturning his pot or the metaphor of the beaver, which points to a shared body of knowledge and encoded meaning that helped Mi'kmaq to survive culturally, spiritually and physically, and find their way as they travelled and lived throughout Mi'kma'ki.

Like all oral histories, these legends can be adapted to local circumstances or be altered to reflect changing times, but this does not negate the inherent value of legends as the earliest recording of the history of Mi'kma'ki—its land and its people—a history and tradition that pre-dates the arrival of other cultures by thousands of years, perhaps more than 10,000 years.

CHAPTER 4
DANCE AS MIRROR

In 1610, the Jesuit missionary, Pierre Biard commented in reference to the Mi'kmaq: "As long as they have anything, they are always celebrating feasts and having songs, dances and speeches.... (Biard 1959: 107). At first glance, his comment might create the image that the Mi'kmaq did nothing but eat, drink and be merry. What Biard was actually documenting was the multi-sensory nature of how people tuned into, mirrored and communicated their experience and knowledge of their environment. Similar to the legends, the dances and songs of the Mi'kmaq encoded information about the environment, sometimes compressing multiple layers of experience and information into a single step or chant. Dances, whether formal or informal infused Mi'kmaw culture, and were inseparable from everyday life.

More than 125 contexts in which dance or gesture occurred within Mi'kmaw culture have been documented (Sable 1996: 228; www.native-dance.org). No doubt there are more. Dance was a means to invoke power and to embody the spirit of animal, plant, enemy or lover. It was used to seduce, cajole and conquer. It was a way to celebrate, give thanks, honour another, dispel grief, prepare for the hunt and war, heal the sick, trade goods, court, celebrate marriage, mark a rite of passage—and to have fun. Even in the "Land of Souls," referred to by Chrestien Le Clercq as the place where people's souls or spirits go following death, there was dancing (Le Clercq 1968: 88). In fact, most everything in Mi'kmaw culture was at one time danced into being. As the dance anthropologist Alan Merriam states, "dance is culture and culture is dance" (Merriam 1964: 17).

The same principles that apply to storytelling apply to dance. Dance is unique because it is kinesthetic—it relies on the body as the main instrument of expression. Yet, the power of dance is the ability to communicate through many channels simultaneously.[1] Dance is multi-sensory, reflective and contextual. It is generally interactive and communal, with a continual exchange occurring both between dancer and onlookers, and/or between dancer and the world of energy in its many forms. It reaffirms values and transmits knowledge through known and patterned movements and rhythms while simultaneously allowing for individual interpretation. Dance can also be spontaneous. As noted by dance anthropologist, Alan Lomax, dance is "the most repetitive, redundant, and formally organized system of body communication in a culture" (Lomax 1968: 223).

Aside from the sensory, dance was inspired by dreams. In Mi'kmaw culture, one entered other realms of existence through dreams, chants and dances. Dreams to the Mi'kmaq were carefully regarded as omens, both good and bad and these dreams were often danced/acted out.

> Micmac dreams frequently contain much culture reference. A recurrent theme is the necessity to master an attacking person or animal; actually the dreamer believes he must conquer the witch or buoin, who has sent the dream; otherwise he will be defeated in a daytime encounter. Impressed by the strength of the Micmac belief in dreams, the early French missionaries led planned attacks on "the bonds that held (them) down in (their) wretchedness." But all they attempted to destroy was not misery. According to a 1607 report, those who had auspicious dreams rose in the middle of the night to hail the omen with song and dance. (Wallis and Wallis 1955: 138-39)

In discussions with Mi'kmaw dancers and Elders, reviews of historical descriptions of dance events, and experiences at powwows (*mawio'mi*, "gathering"), dance was never depicted as an abstraction, but as an entering further into the world, a joining with the rhythms of the drum, the earth, ancestors and the other dancers. Vivian Basque, a Mi'kmaw dancer from Eskasoni, spoke of dancing as a way to keep in balance with nature or a way to connect oneself with nature:

> People used dance to call out spirits. They used to be able to enter another world or different states of mind to seek answers and communicate with each other telekinetically [telepathically?]. There was a time when medicine men would be able to travel into different levels of consciousness and have control over their powers. (Basque, interview 1990)

Mi'kmaq have described dance as both a form of prayer and a form of gift-giving. The word *alasutmaqney* means "a prayer in a form which can be a dance." At powwows held in a number of Mi'kmaw communities throughout the Maritimes, the dancing is said to give energy to the drummers and vice-versa. At the end of every powwow is a gift-giving ceremony in thanks to all the people who helped host the event.

From looking at the language it can be seen that the Mi'kmaq had both formal dances and informal dances. *Nskawaqn* is a serious and ordered dance. Silas Rand referred to the *nskawaqn* as the "mystical dances." *Amalkay* means "any old way to dance, just move your body," or an "ordinary dance." There were distinct male and female dances and dance steps. There were also chief dancers, *nuta'lukwet*, dancers recognized for their skill and power in the dance. With colonization, another distinction arose between "dancing like an Indian," *l'nu'tesin*, and "dancing like a white man," *aklasie'wtesin*.

Dance is both an expression of "self" ("you dance when you're happy, you dance when you're sad" said Vivian Basque) (1990), and a means to unify people into a circle of energy or power. Radcliffe-Brown, in speaking of his experience among the Adaman Islanders, described dancing as, "a means of uniting individuals into a harmonious whole and at the same time making them actually and intensely experience their relation to that unity of which they are members." Similarly, Richard Waterman describes dance as a "force for social cohesion, and as a means to achieve cultural continuity" (Radcliffe-Brown 1964: 283; Waterman in LeBlanc 1995).

Dance was also way to work with and channel emotions, and to dispel or prevent being overwhelmed by forces that interfered with a person's ability to communicate and survive. One account told of Mi'kmaq dancing when famine struck; another danced to bring himself back from his grief over the loss of his wife and child. Le Clercq gave an account of the dance and song done by the husband following the recital of a funeral oration for the man's wife and child:

> Our Koucdeaoüi had no sooner received these public approvals, than he set himself to dancing his very best and to chanting some songs of war and the chase, in order to testify to the assembly that he had banished from his heart all the regret, grief, and sadness he had previously felt. (Le Clercq 1968: 187)

There are many definitions of dance, each emphasizing different functions or characteristics of it. Many of these define dance in terms of what it is not: human versus animal (Hanna 1988: 37), non-utilitarian versus utilitarian (Royce 1977: 1, 5), mythic versus scientific[2] (Langer 1983: 38), and extraordinary movement versus ordinary movement (Hanna 1988: 37).[3] Although the Mi'kmaq had their own set of dance movements unique to their culture, the essence of Mi'kmaw dance, like the language and the stories, was a means to mediate among the various forces at play in the world and bring them into a knowable or visible relationship for those gathered.

Dances were also exchanged between different cultural groups, making it a cross-cultural form of communication and a way to honour and give to others. Dances were seen as a form of gift-giving (e.g., the "pine needle" dance was said to be "given" to the Mi'kmaq by the Passamaquoddy (Smith n.d.).[4] Vaughen Doucette of Eskasoni was told that the snake dance (or serpent dance) was given to the Mohawk at the Great Council of 1749. In return, the Mohawk gave the Mi'kmaq the thanksgiving chant, "*I'ko*" (Doucette, personal communication 1995). Interestingly, dance and gesture were perhaps the first way foreigners saw and communicated with the Mi'kmaq. In 1534, Jacques Cartier wrote the following account while on the north shore of the Bay of Chaleur:

Upon one of the fleets reaching this point, there sprang out and landed a large number of Indians, who set up a great clamour and made frequent signs to us to come on shore, holding up to us some furs on sticks. But as we were only one boat we did not care to go so we rowed towards the other fleet which was on the water. And they [on shore] seeing we were rowing away, made ready two of their largest canoes in order to follow us. These were joined by five more of those that were coming from the sea, and all came after our long-boat, dancing and showing many signs of joy, and of their desire to be friends, saying to us in their language: *Napou tou daman asurtat*, and other words we did not understand.... The next day [Tuesday, July 7] some of these Indians came in nine canoes to the point at the mouth of the cove. As soon as they saw us they began to run away, making signs to us that they had come to barter with us; and held up some furs of small value, with which they clothe themselves. We likewise made signs to them that we wished them no harm, and sent two men on shore, to offer them some knives and other iron goods, and a red cap to give to their chief. Seeing this, they sent on shore part of their people with some of their furs; and the two parties traded together. The savages showed a marvellously great pleasure in possessing and obtaining these ironwares and other commodities, dancing and going through many ceremonies, and throwing salt water over their heads with their hands. (Biggar 1924: 49-50, 52-53)

The gestures and dances allowed for communication across cultures, not the words spoken to Cartier and his crew by the Mi'kmaq. These movements indicated to Cartier invitation, friendship, a desire to trade and joy and pleasure once trade took place.

Dance also occurs in different contexts, which also can be significant to the meaning and the importance it plays in communicating information. Context includes the culture in which it occurs, the space in which it takes place, the people who participate both as dancers and on-lookers, the time it is held and the general purpose for which the dance is being done.[5]

In Mi'kmaw culture, dances generally occurred within a communal setting where everyone present witnessed and/or partook of the dance experience. These dances were accompanied by songs and chants, along with the beating of a birch bark drum or with *ji'kmaqn* (split ash splints bound together at one end, and hit against one's knee or the palm of the hand to make a sound—figure 19).

One of the most complete descriptions of a dance event was documented by Pierre Antoine Maillard in 1758.

That particular event was held to honour the departure of a visiting envoy. Once the food was prepared the host called the men of the village

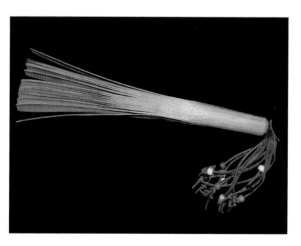

Figure 19
A ji'kmaqn. Photo by Trudy Sable

to the feast, excluding women, children and young men who had not yet made their first kill. The men then entered the wigwam of the host carrying their *oorakins*[6] (birchbark dishes), were seated in accordance with levels of honour, and served the meat, accompanied by another *oorakin* filled with oil. Following the feast, the wives of the men entered to take away the bowls. The women then departed and a period of profound silence and pipe smoking ensued. Eventually, as if rising from a trance, the guest of honour rose, commenced a long oration several hours in duration in honour of the host, regaling the host's ancestors as far back as his great, great, great grandfather, and the prowess and skill of the host in the hunt. The speech, as always, was filled with eloquent metaphors. Finishing this speech of thanksgiving, a second, younger man arose to thank the master of the feast, but first commended the speech of the first orator. Following his speech of thanks, the younger man then began to dance the *neskawet*. *Neskawet* is a verb meaning to "sing with gestures and responses" and is a more formalized style of dancing.

Then quitting his place, and advancing in cadence, he takes the master of the treat by the hand, saying, "All the praises my tongue is about to utter, have thee for their object. All the steps I am going to take, as I dance lengthwise and breadthwise in thy cabbin are to prove to thee the gaiety of my heart, and my gratitude. Courage my friends, keep time with your motions and voice to my song and dance." With this he begins, and proceeds in his *Netchkawet*, that is, advancing with his body strait erect, in measured steps, with his arms a-kimbo. Then he delivers his words, singing and trembling with his whole body, looking before and on each side of him with a steady countenance, sometimes moving with a slow grave pace, and then again with a quick and brisk one. The syllables he articulates the most distinctly are, Ywhannah, Owanna, Haywanna, yo! ha! yo! ha! and when he makes a pause he looks full at the company, as much as to demand their chorus to the word Heh! which he pronounces with great emphasis. As he is singing and dancing they often repeat the word Heh! fetched up from the depth of their throat; and when he makes his pause, they cry aloud in chorus, Hah!

After this prelude, the person who had sung and danced recovers his breath and spirits a little, and begins his harangue in praise of the maker of the feast.... Then he takes them all by the hand and begins his dance again: and sometimes this first dance is carried to a pitch of madness. At the end of it he kisses his hand, by way of salute to all the company; after which he goes quietly to his place again. Then another gets up to acquit himself of the same duty, and so do successively all the others in the cabbin, to the very last man. This ceremony of thanksgiving being over by the men, the girls and women come in, with the oldest at the head of them, who carries in her left hand a great piece of birch bark of the hardest, upon which she strikes as it were a drum; and to that dull sound which the bark returns, they all dance, spinning round on their heels, quivering, with one hand lifted, the other down: other notes they have none, but a guttural loud aspiration of the word Heh! Heh! as often as the old female savage strikes her bark-drum. As soon as she ceases striking, they set up a general cry, expressed by Yah! Then, if their dance is approved they begin it again.... (Maillard 1758: 12-15)[7]

The women then dance to exhaustion but, before departing, the eldest extols the strength of women in yet another lengthy oration filled with metaphors and passionate descriptions of women's feats.

The reflective or mirroring nature of dance can also be seen. The dancer entreats those assembled to, "Keep time with my motion and voice to my song and dance." They are being entreated to "tune in" to his rhythm, his words, his energy. The profound silences, and pipe-smoking, the arising from a trance-like state, are also a means of "tuning in" and acknowledging the mutual space in which they all co-exist.[8]

As seen in this example, dance embodied the values inherent in the culture. Like everything else it was and is considered a form of gift giving and exchange. Individuality and communality are also seen in dances. It was at once an expression and acknowledgement of one's individual being and ancestry, and a sharing and exchange with one's community.

Dance competitions also embodied this quality. The late Joey Gould of Apaqtnekek (Afton) First Nation, originally from We'koqma'q (Whycocomagh), described dance competitions that used to occur at St. Anne's Day celebrations:

> On Sunday evenings, there would be a Ko'jua competition to see who would "take the Ko'jua home." Dancers would get up on the stage and start competing to see who was the best dancer, and who could take the Ko'jua home to their reserve. Joey said: "As soon as I heard people hollering at the dancers, "It's going to Eskasoni, It's going to Membertou, It's going to Nyanza—nobody would mention Whycocomagh—then that would give me a feeling to get up there.... I'm going to beat them all. As soon as I started dancing my style, I could hear my people from my reserve and even from Eskasoni where I have relatives, they would start calling, "it's going to Whycocomagh. It's going to Whycocomagh." You'd dance the Ko'jua until you just tired out and can't dance anymore. I think that's the way you'd compete. You started and go out on stage until you can't dance anymore. (Gould, interview 1991)

Dance competitions are commonly featured in legends. One *puoin* would dance against another to prove their power. Dance was used to invoke power, transform shapes and conquer enemies.

Like stories, dance contexts changed over time as new cultural influences were adopted. The Catholic church played a primary role in discouraging dance, particularly the shamanic and ritualistic ones having to do with healing, birth and death. However, like stories, certain cultural themes and values endured. *Pestie'wa'taqatimk,* or the "Naming Ceremony," was done by Mi'kmaq on designated days throughout the Christmas season to celebrate certain saints' names. This was a Christian ceremony most likely introduced by the missionaries. Dr. Marie Battiste said that prior to centralization when Mi'kmaq lived in smaller settlements (centralization occurred in the 1940s and 1950s) people would go from house to house giving gifts to people who had the same name of a particular saint being celebrated that day (i.e., Louie (Louis), Newell (Noel), Anne, etc.). After offering a gift to the householder whose name was being celebrated, the visiting group would enter the house and dance the *ko'jua*, a celebration dance. This dance was an offering in exchange for the food the host would then provide (Battiste, interview 1992).

ENVIRONMENTAL KNOWLEDGE EMBEDDED IN DANCE

The serpent dance is a Mi'kmaw dance that vividly exemplifies ways of connecting with and coming to know the world. This dance is sometimes done at powwows. During the serpent dance people, primarily children, line up behind the lead dancer placing their hands on one another's shoulders, or holding hands if they could not hang on. The dance step is like a heavy-footed, careful run, with a slight side to side motion. As the chanting begins, the head dancer leads the line in serpentine movement, winding back and forth until reaching the opposite side of the dance circle. Once on the opposite side, the beat slows and the head dancer remains in the centre of the circle of dancers as they spiral or coil inward around him. Once the coiling is complete, the tempo increases and the head dancer comes out through the circle of dancers, and again begins winding back and forth with the line in tow back to the other side, still counter clockwise. Again the tempo slowed as a second coiling occurs. After the head man exits the second coiling, he again proceeds in a less winding fashion around the circle, counter clockwise, picking up the tempo a bit and finishing the dance when the drum and chanting ceases.

How far back this dance goes among the Mi'kmaq is difficult to say. As previously mentioned, Vaughen Doucette of Eskasoni was told that the snake dance (or serpent dance) was given to the Mohawk at the Great Council of 1749 (Doucette, personal communication 1996). In another account, the dance was performed in 1860 by Maliseet for the Prince of Wales at Government House in Saint John, New Brunswick (Whitehead, personal communication 1996), a sketch of which exists in the collections at the McCord Museum in Montreal.

Versions of the snake song and dance have been documented among the Penobscots by Frank Speck, William Mechling and Walter J. Fewkes, around the turn of the 20th century. Stansbury Hagar documented the *jujijuia'tijik* (Hagar writes it *choogichoo yajik*), or the "serpent dance" (literally, "they are making like a serpent") as remembered by Newell Glode in 1895. Glode reported that very few members of his community remembered the dance, and that he himself had forgotten the song that accompanied the dance. At the time, Newell Glode was seventy-three years old (Hagar 1895: 37).

The word, *jujijuia'tijik*, according to Margaret Johnson means "acting like a snake." She explained the word *jujij* refers to things that crawl on the ground, like snakes, lizards or spiders. This is in keeping with Hagar who quotes Rand's definition for it being a general term for reptiles. Hagar also mentions that, despite this definition, several Mi'kmaq assured him it designated the rattlesnake.

Nicholas Smith cites Jack Solaman, a Maliseet from Tobique Point, NB, as using the word "*al-la-de-gee-eh* (*alatejiey*: literally "I am slithering around") in a singing of the "Snake Song" in 1915. This word was translated by Peter Paul as "moves like a snake." Rather, the word literally means that "it has the motion of a snake." In Mi'kmaw this word is *alatejiey*, which is translated as "slithering around" or "the movement the snake makes," which is similar to Margaret Johnson's definition. The actual word for snake in Mi'kmaw is *mte'skm* (Johnson interview 1995; Smith n.d.).

The dance, as documented by Hagar, was done by both men and women who formed a circle around a chanter and lead dancer.

> The circle of dancers moved first to the right three times around the head man. The dancers then turned their backs to the head man and repeated the revolution three times; next the two sets turned their backs to one another and again moved thrice around the circle; finally, in the same position, they reversed the direction of the motion and move backward around the circle three times. This figure was thus completed in four positions and twelve revolutions, and, according to Newell Glode, signifies the rattlesnake waking from his winter sleep. (Hagar 1895: 37)

> The head man now left the circle through the space made for him, simulating a serpent coming from its hole; he led the dancers around the field, making many snake-like twistings and turnings. In one hand he held a horn filled with shot or small pebbles; with this he rattled the time for the step and the song of the other dancers. After they had advanced some distance the last dancer remained stationary and the others moved around the leader in a constantly narrowing circle until all were closely coiled around him. The head man then reversed the direction of the motion and the dancers came out of the circle in line as before. This represented the coiling and uncoiling of the rattlesnake. (ibid.)

Hagar describes the line twisting and turning around the field, coiling and uncoiling around the headman. This was done three times in order that "the rattlesnake can shed its skin." Finally, the head man leads the dancers back toward the centre, whereby the dancers dropped away "at regular intervals" leaving only the headman alone to end the chanting and dancing.

Hagar also reports that:

> the Micmacs assert that the traditional object of the dance was to obtain the poison of the serpent for medicinal use, and that at one time long ago their ancestors used to dance it so much that nearly all of them were turned to serpents" (Hagar 1895: 38).

Another legend speaks of a community turning into toads from ceaseless dancing. Perhaps the dancers invoked the spirit of the snake or toad so much, that it overpowered their own human form. The snake dance was supposedly danced to coincide with exuviation, or the moulting of the skin, and was also done at the election of a chief. This may have been done to empower the new chief or give him strong medicine for protection (ibid.).

Additionally, Hagar alludes to the dance being associated with the Pleiades, known, as he says, in the Mayan culture as the "rattle asterism," and states that "by the Micmacs own interpretation of the dance that it refers to the seasons...." (Hagar 1895: 38). Frank Speck notes that among the Penobscot, Newell Lion had heard that "the dance anciently represented the movements of a serpent (constellation) in the sky. But he could give no further connection with it" (Speck 1976: 284).

In a separate account written a year later, Hagar loosely infers that the dance was done in conjunction with the collecting of a medicinal plant called "*meteteskewey*,"[9] translated as "the rattling plant." According to Margaret Johnson *metetesk* means "rattling," and the whole word connotes something that rattles (Johnson, interview 1996). Similarly, Hagar writes that "its three leaves strike each other constantly with a sound like that of the rattlesnake." Hagar's informants stated that the plant "resembles a wild turnip."

> It stands about knee high, with leaves about eight inches long, like those of the poplar. Its root is the size of one's fist, and the stalk is surrounded by numerous brownish yellowish balls as large as buckshot. Others describe the plant as being much smaller. (Hagar 1896: 175-76)

In Hagar's second account, he reports of the appropriate ritual for collecting the plant by an individual.

> To find the plant, one must first hear the bird called *cooasoonech* [*kua'sunej*: "dwelling in old logs"] singing in an interval in the forest, otherwise the plant is invisible. This bird is brown and very small, but is chose chief of all birds because he is quickest and can hide in the smallest holes. He is sometimes called *booin*, "the magician," from his aptitude for quick disappearance and his ability to fly through fire without being injured. When he sings, one should follow him at once, although ... he often leads one on and on through the forest depths, leaving him lost and forlorn. But the fortunate one will at last hear the rattling of leaves of the magic plant as he approaches it, and then the plant itself will soon be seen. He must now gather thirty sticks and lay them in a pile near the plant. Next he must induce a girl, the more beautiful the better, to accompany him to the plant. Under circumstances of the greatest

temptation, both must have no wish save to obtain the medicine or the plant will disappear. Now the plant is inhabited by the spirit of the rattlesnake, which comes forth as they near the plant, and circles around it. The man must pick up the serpent, which will then disappear without harming him. These tests of perseverance, self-control, and courage are all I have heard, but there may be others. The plant must be divided in four portions, of which three may be taken, but one must be left standing. The three parts are scraped and steeped and a portion worn about the person. Some say that, divided in seven parts, this medicine will cure seven diseases, but the great majority believe that it will cure any disease and gratify any wish. It is held to be especially potent as a love-compeller. No woman can resist it.... [Hagar 1896: 175-76]

Hagar finally mentions that

The rattlesnake which accompanies the plant brings it at once into touch with the mysteries in all parts of the globe. The same species is associated by the Micmacs with a dance which they used to perform only at night. This dance was mystical in a marked degree and was connected with the Pleiades. (Hagar 1896: 176).[10]

It is possible that the dance diffused to the Maritime from other areas, was "given" to the Mi'kmaq by another people, or, less likely, done in honour of the Maritime garter snake or some other snake(s) whose shedding of skin marked the changing of seasons. More likely, the dance is in reference to the *jipijka'm*, or horned serpent prevalent in Mi'kmaw legends. Parsons also surmises that the serpent dance, as reported by Hagar, is in reference to the *jipijka'm*, and not the rattlesnake (Parsons 1925: 60).

It may be speculated that the *jipijka'm* was the keeper or spirit of medicine (Sable 1998). The association of serpents and medicine is seen in numerous cultures throughout the world. The *jipijka'm* is a powerful symbol in Mi'kmaw legends. It lives beneath the earth or water, and its horns, one red and one yellow, were used for personal power particularly by *puoinaq*.

In their snake shapes, they travel about under the earth, swimming through the layers of rock; the ground trembles as they pass. Sometimes they come up to the Earth World, and carve great ruts in the land as they move across it. They live as humans in the World Beneath the Water. Sleeping, they can seem mountains. All *jipijka'm* have one red horn, one yellow horn; these horns are Power objects, and stories about their use are known westward all the way across Northern North America and through the centuries, back into Northern Asia. (Whitehead 1988: 4)

Margaret Johnson defined *jipijka'm* as like a crocodile, or a big *jujij* that lives in swamps. Isabelle Googoo Morris told Parsons that *jipijka'm* "live in big swamps rooting in trees.... Every chibeshkam' [*jipijka'm*] has a big red or yellow horn. This horn has magical application and such a horn was part of the outfit of a witch (*bu'owin*) [S/F *puoin*]" (Parsons 1925: 60). Hagar describes it as "a horned dragon, sometimes no larger than a worm, sometimes larger than the largest serpent.... He inhabits lakes, and is still sometimes seen" (Hagar 1896: 170). Perhaps, as well, its serpentine aspect was connected with the physical snake as a marker of seasonal change. This, in turn, might provide an indicator for the proper time to pick medicine.

In a sense, there are three levels of meaning, all inseparable. On the external level, the rattling of the horn, filled with pebbles, mimicked or reflected the rattling or tapping sound of the plant. The chant that accompanied the dance, Margaret Johnson suggested, may have mimicked the beat of the rattling leaves or stalk—*metetesk, metetesk, metetesk*—onomatopoeically. This plant, in turn would be powerful medicine for the people, if properly respected. On another level, the dance and chant most likely was part of becoming, awakening, honouring and possibly testing the energy of the *jipijka'm*, the essence of the medicinal plant. The plant itself may have mirrored the features of the *jipijka'm* in both appearance and the sound it made. The sound and the rhythm of the dance embodied the essence or nature of the *jipijka'm*, which inseparable from the medicine. The third level of the dance may have had to do with "turning over" the seasons, as also seen in legends, and connected in some way with the stars (Sable 1998).

Figure 20
The late Dr. Margaret Johnson, who contributed greatly to the research in this book, and author Trudy Sable at Chapel Island (Potlotek), NS. Photo courtesy Trudy Sable.

The first level has already been discussed in terms of the appearance, sound and potency of *meteteskewey* as a medicine used by the Mi'kmaq. The second, or inner level, has to do with the nature or essence of the medicine in the form of the *jipijka'm*. The nature of *jipijka'maq* is like medicine in that medicine can poison or cure someone depending on proper use. This in turn is similar to the power attributed to *puoinaq* (powerful and often feared medicinemen/shamans) who, in legends, often had *jipijka'maq* as allies. *Puoinaq* were known to exercise both healing and destructive powers. Perhaps the two horns of the *jipijka'm* represent red and yellow ochre, both minerals collected for paint, and used as power substances. Red ochre in particular was thought to have powerful magical properties and was used as a medicine (Mechling 1958: 242; Sable 1998). This may have been true of yellow ochre as well.[11]

The third level has to do with the dance being performed to mark or effect the changing of the seasons. There are two particular legends which talk about the turning over of seasons, both in connection with medicines.

The first is "Djenu and Kitpusiagana" (Jenu and Kitpusiaqnaw) told by Peter Ginnish of Burnt Church, New Brunswick.

> Djenu was the strongest man in the world. When he became angry, he grew. He is under the ground, alive, to this day. Djenu lies as he was buried. Kitpusiagana was another strong man, not dead, who is buried in the ground. One who goes where he is buried obtains medicine. Twelve men go every three months and turn him from one side to the other, from his face onto his back, and then onto his face again. Everything grows above him. When you pull something that is growing above him, you obtain good medicine. A limb of a tree or a bush which grows there will cure anything. He is buried in the east, Djenu still further to the east. They fought, and many people died because of the noise they made. I do not know on what month or day he is turned. (Wallis and Wallis 1955: 343)

Similar to the serpent dance, there are three revolutions done in each direction, four times, possibly for the three months of each of the four seasons. As seen in the this dance and the serpent dance, the dancers first face one another, then face outwards, similar to Kitpusiaqnaw being turned from his face onto his back, and then onto his face again. Again we see that good medicine can be procured by properly turning Kitpusiaqnaw over at the appropriate times. Jenu, is said to have a heart of ice, and a scream that can kill. Possibly Jenu is connected with storms, easterly winds or cold weather, that prevents medicines from growing so that people cannot be healed. Or, possibly Jenu is the poisonous potential of the plant.

The second legend is of Kulpujot ("rolled over by handspikes") (Rand 1971: 234).

> [Kulpujot is] an old man who dwells in solitude broken only by occasional visitors.... He is without bones, and his corpulence is so great that he lies upon the ground in one position, unable to move. Twice a year, in spring and Autumn, he is turned over by visitors armed with handspikes, hence his name. Kulpujot, in this story lives far to the south. When he lies facing the north, his warm breath produces those balmy southern zephyrs which bring with them the song of birds, the perfume of flowers, and the wealth of summer vegetation. When he is turned towards the south, the birds and flowers follow, and the icy northern winds resume their sway. (Hagar in Hoffman 1954: 248-49)

Kulpujot supposedly could grant any wish desired to those who turned him over. One of seven men who venture to visit him asked to live with him and serve him water and tend his fire forever. The man had his wish literally granted and was subsequently turned into a cedar tree (Hagar in Hoffman 1954: 248-49). Again, the cedar has medicinal uses, as well as being able to live in swampy, watery areas. Nevertheless, the turning of the seasons seems to correlate with collecting medicines.

Joel Denny of Eskasoni described the giving out of medicine to the people by the medicine man. This involved a medicine song and a dance whereby the people danced up to the medicine man, received their medicine, and then returned to their place after dancing a more vigorous version of the dance (Denny interview 1993).[12]

These examples were provided to illustrate the power of dance as a medium for mirroring and communicating knowledge that was important to survival of the Mi'kmaq as well as a means to embody and honour their relationship with their environment. This dance is yet another teaching of respect for the powers at play, some of which can kill you, such as picking the wrong medicine. It also teaches of the seasons, the directions, the stars, the nature of reptiles, the bird that leads one to the medicine and values of respect and care needed in collecting plants. (There is also a possible association with the Pleiades but more research is needed.) Offerings to the four directions were made in the dance, acknowledging the gift of the medicine. Properly approached, the medicine will be good and strengthen the people. From the study of one dance, or one plant, a whole web of relationships and information about the world comes into being.

CHAPTER 5
SONGS AND CHANTS AS MIRRORS

It has been said since ancient times that the nature of reality is much closer to music than to a machine (Capra in Berendt 1991: xii).

The Micmac learned songs from birds.... They learned especially from the wild turkey and the sea gull. "Ka ka ka kwi't," sings the wild turkey to herald a storm. Gulls which fly around together and herald a storm or a high wind for the following day furnished inspiration to the Micmac composer. The gull sings "Ka'ni! Ka'ni! Ka'niak! Ka'niak!" three times, then flies away. One old Indian listened to the gull until it had finished its song. Thus he learned its song, and said to the others: "If you people care to dance to it, dance. If not, then merely listen to me." He then took a stick and beat time. But as he sang he wanted to put some words to the tune. He was thinking about a woman who was hunting for something, and accordingly he sang about this. (John Newell in Wallis and Wallis 1955: 119)

Discussions in the previous chapters are exemplified in this description of the creation of a song. Listening to the sounds of the birds heralding a storm, a song is created, mirroring the call and rhythm of the birds who, in turn, are mirroring the weather. The birds give people their song. These sounds are brought into the human realm of experience further by associating them with a woman hunting for something.[1] The singers then invite those around them to dance, or listen, depending on their inclination, thus giving them the song. Weather, birds, song creation, dance, story and woman hunting—in one song (Sable 1996: 258; 2006: 262).

Joachim Berendt, a student of physics and a renowned jazz musician, asserts that the sciences have neglected the world of sound. He asks: "Why is it that science has failed to emphasize the role of sound in the world?" In Berendt's opinion, science has overemphasized the visual to the detriment of the other senses (Berendt 1991: 9). Berendt's book, *The World is Sound: Nada Brahma*, describes in both musical and scientific terms, how everything in the world is sound down to the vibration of molecules. Berendt notes that animals are so subtly tuned to one another's rhythms based on vibrations, they are able to move in synchronization, such as schools of fish and flocks of birds.[2]

There are many songs in Mi'kmaw culture that mirror the sounds, rhythms and features of nature, both with words and onomatopoeically. There is a snake song, a wind song, a pine needle song, an eagle song, a loon song, a toad song and many others. For Mi'kmaq, both knowledge and values were embedded within the song texts and the musical structure that were passed from generation to generation. Vaughen Doucette, a traditionalist drummer and chanter from Eskasoni explained:

> When you are teaching songs you're teaching not just the song, but a lot of etiquette, talking about power and how to channel power. There's a lot of spin off from that regarding your place in the world. Not just teaching a song, but the dos or don'ts. You have to make preparations, invite spirits to come, and the song is a prayer for all the people who need help. A lot of social issues are dealt with the whole time. (Doucette, personal communication 1996)

In many cultures, breath was and is associated with life force, inseparable from the life energy that infuses all of creation. For instance, when the Maori invite guests from other cultures, they are formally chanted into the Marae (spiritual meeting house). The Maori hosts then form a semi-circle while the guests move left to right, person to person, touching forehead to forehead and nose to nose to exchange breath with each other.[3]

Breath, sound and word are part of a continuum and a way to channel energy, shape reality and to literally "tune in." According to Isabelle Knockwood, word is your spirit, and some stories were said to have the power to change the weather (Knockwood 1992: 14; Hagar in Hoffman 1955: 250). The Aborigine song lines of Australia are an example of the world being sung into being, not just once, but continuously, reaffirming and reconnecting a person with the landscape and their place and identity within it. Songs, and song parts are associated with the features of the land.

Music and songs of specific cultures express(ed) fundamental world view, both in their musical structure and in the song texts.[4] Each culture has its own rhythms and shaping of sounds that reflect its particular cultural message.[5] Among the Gros Ventre (a historically Algonquian-speaking people in north central Montana), Orin Hatton demonstrates how a war expedition song is a paradigm of their creation myth. The song embodies two of the Gros Ventre's main values of mastery and resourcefulness, but both within the context of humility and respect for the creative forces of the world (Hatton 1990: 52).[6]

Parsons recorded the tale, "Brings Back Animals," told to her in 1925 by Isabelle Googoo Morris, as heard from Bessie Kremo Morris of Sydney, Nova Scotia. The following is just one portion of the story:

...The man in the wigwam began to beat on bark and to sing. He said, "I am singing of the animals, all the animals (*waisisk*) to come alive, to come back to life, from all these parts, wings, heads, feet, that have been thrown away." He sang: *negane'sung'ul besikwia g'ul*—what belongs to my feet I am losing [moccasins]. He stopped singing at daybreak. In the morning he said to the visitors, "That is my work every night. I don't like to see people waste any part of the animals. They should save everything, they should save eel skins and other parts...." They went down to the shore. He said, "Do you want to see the fish come?" He took out a shell whistle. The bottom was very clear. They could see all kinds of fish. "These are my fish," he said. "They come from all those parts people throw away on the shore. I sing for them and they come back." (Parsons 1925: 72-73)

The lesson of wastefulness and respect for environment is an obvious message in this story. In Mi'kmaw culture almost every part of an animal, plant or fish was put to use. William Mechling noted that Mi'kmaq believed "one of the most frequent causes of misfortune was the wasteful slaughter of game" (Mechling 1958: 198). The story also conveys the qualities of gentleness and caring, as well as exertion in working hard all night to make sure the animals come back. This account contains the reminder that the bones of animals and fish should be placed in the appropriate places so that these animals would return in their physical forms. On the most inner level of meaning, it affirms the connection between one's own life and that of the animals. In Parson's transcription, the man blows into his shell whistle, using it as the physical medium to call back the fish with breath. Song, an extension of one's own breath, is connected with the life of animals (Sable 1996: 262).

This quality of connectedness, mirroring and bringing and singing creation into being, is not just in relation to animals. Frank Speck's account of a group of Penobscot men canoeing across the Penobscot Bay in Maine in the Eastern United States, illustrates the joining of chant with the rhythm of the waves:

The magic power of song syllables was thought to have a quieting influence upon the forces causing rough water, and also to strengthen the canoe men. A number of years ago an informant (Charlie Daylight Mitchell) was crossing from Deer Island to Eagle Island in Penobscot Bay during a heavy sea. He was in a small canoe in the company of an old man who chanted ... all the way across. The singer tempered his voice to follow the pitching of the canoe as it mounted wave after wave. He said that the boat rode the waves much more easily while the old man was singing. (Speck 1940: 167)

Speck goes on to transcribe a musical score for the chant, noting: "The meaningless syllables are repeated over and over again, *kwe ha' yu we, ha'yu we hi' / Kwe ho' you we, ho' yu we*" (ibid.).

Similarly, Nicholas Smith noted that Penobscot men developed "cadence songs" to accompany the rhythmical routines of walking and paddling. The song leader could easily vary the pace of the song to meet the needs of the travellers by changing the tempo (Smith n.d.).

The monosyllables sung or chanted by Mi'kmaq and transcribed by early chroniclers—ho, hé, ha, houen, etc.—were commonly termed nonsense or meaningless syllables.[7] Although these syllables cannot be given definitions, they were far from meaningless. Hatton's work among the Gros Ventre, as well as research in Eastern cultures (e.g., Tibet), shows us that these monosyllabic chants were perhaps the most esoteric level of rhythm, sound and music. They are encapsulations of the essence of life, and the way a person communicated directly with the world. As Vaughen Doucette expressed it, "chanting is more spiritual, singing can be spiritual, but done with words. Chanting with emotion is spiritual" (Doucette, personal communication 1996). Song syllables are perhaps the most fundamental level of "tuning into" the sounds, vibrations and rhythms of the world as seen in the Penobscot cadence songs. Among the Gros Ventre, singing was thought to degenerate "when songs were composed and used for 'trivial things'" (Cooper 1957: 79). Interestingly, it was the birds who were thought to maintain the dignity of the songs.[8]

> Birds and other creatures are conceived as having kept the power and dignity of their singing intact. An important aspect of the quest for supernatural power is retrieving the power of the moment of creation from the creatures that attended and celebrated the reformation of earth. (Hatton 1990: 53)

An account of learning songs from owls might be viewed in this light. (Owls portended death if they appeared in the early morning.)

> Many songs were learned from owls. One can almost understand their language. They speak words distinctly, "wi ya," (long drawn out) is said by the owl, plainly, when he has finished his song. After this the owl leaves and another takes his place, and sings "wuk'wa ha, wuk'wa ha." After he has finished he goes off to one side, sits by himself, and gives a "hu'a" (rising inflection on first syllable, falling inflection on the second). Then a "wa'hi! Wa hi!" and at each syllable nods his head. A Micmac learned a song from the owl. He was camping in the woods. The owl smelled the fire and came close to it. After a while he heard someone singing "Ru! Ru! Ru! Ru! Ru! Ru! Ru! Hua'wa! Hua'wa!" finishing with "Hi'a! Hi'a!, Hi'a!" Soon another owl sang; and then another. The owl commenced with high notes, in a very sweet voice, and ended with "hu'a!," a deep guttural. It sounded as though he was choking. After this, three or four stand in chorus, the refrain of one answering the syllables of the other. The man said, "I have his song now, and will sing it here after as a *neskawe't* [*sic*; proper spelling is *neskawet*: "greeting song"]. (Wallis and Wallis 1955: 119)

This song mirrors the more formal Mi'kmaw chants and dances done in greeting and praise, and for other important events. *Neskawet* is a verb meaning, "he dances a particular dance and perhaps sings with gestures and responses." Furthermore, if someone is either singing or singing and dancing you will hear the group responding with "ahe, ahe." *Neskawet* is also defined as a more serious and ordered dancing, done for a particular occasion such as the election of a new chief. *Neskawintu* (verb) means, "I sing the *neskawet* song." *Nskawaqn*, means a "Mi'kmaw song, this song," and is simply the noun form (Pacifique's Grammar n.d.: 253, 256; Hewson, personal communication 1995/1996; revised by Francis 2011).

Wilfred Prosper explained that *nskawaqn* tells a tale. When asked what it told a tale about he replied:

> I don't really know cause there's no words to it. [Wilfred sings "*I'ko*"] What's that mean? It means nothing. But just the way it's performed, I guess that means something.... Why do they say that they put on a dance for the chief ... after he was selected in 1919? And after it was all over they went into the wigwam and they put on this feast dance. So, what they did meant something to fit the occasion. Certainly not the words. The gestures, maybe.... If it was done for a marriage ceremony, it was done differently. Not much different, it's probably the same song, but they probably did something a little different. (Johnson and Prosper 1992)

The references to *nskawaqn* (usually referred to in its verb form, *neskawet*, by chroniclers of the Mi'kmaq) in historical texts indicate its performance was done within formal contexts. Parsons's account refers to *neskawet* as a "war medicine dance." The late Annie Battiste, an Elder from Chapel Island, also discussed it as "war dancing" (Annie Battiste, interview 1992). Rand defines it as "mystic dance":

Figure 20
Bessie Prosper and photo of her late husband Wilfred Prosper. Both were indespensible to the research for this book. Photo by Trudy Sable.

> At the proper time a chief comes out of a camp, sings a singular tune, dances a singular step and is responded to by a singular grunt from the assembled crowd. They assert that during the ceremony the body of the dancer is impervious to a musket ball. (Rand 1850: xxxi)

In Maillard's account, a *nskawaqn* was done honouring the host of a feast given for a visiting envoy.

> The syllables he articulates the most distinctly are, *Ywhannah, Owanna, Haywanna, yo! ha! ho! ha!* and when he makes a pause he looks full at the company, as much as to demand their chorus to the word *Heh!* Which he pronounces with great emphasis. As he is singing and dancing they often repeat the word *Heh!* Fetched up from the depth of their throat; and when he makes his pause, they cry aloud in chorus, *Hah!* (Maillard 1758: 13)

Elder Dr. Margaret Johnson (1913-2010) once demonstrated the chant and dance done by the members of the Grand Council during St. Ann's Day on Chapel Island in the 1920s.[9] This dance, also documented by Parsons in 1926, was done to the "*I'ko*," ("welcome chant"). Each chief or captain would get up in turn and chant while doing particular dance steps—hands held behind their back, body inclined forward. They would dance around the circle of chiefs and captains, anti-sun wise according to Parsons. At the point in the song where he chanted "*neh*" the dancer would be in front of another chief and accent their movement in greeting. At the end of each round, those assembled would shout "*eh*" (Johnson, personal communication 1995; Parsons 1926: 469-70).[10]

The inference might be made that if *neskawet* is singing a song accompanied by gestures in a more formal or ordered style, then the owl song was also considered in a more elevated category of song than other songs (i.e., songs sung for fun or everyday news). This might be similar to Hatton's statement that birds kept the dignity and power of singing intact. This would also infer a formal code of conduct in relating to birds and other animate beings.

Song was a way to channel energy and use it appropriately. Hatton notes the importance of mentally captivating people with song:

> In sacred songs the all-important thing was the ability to put the proper feeling into the singing, "to get the hearers" to concentrate ears, eyes, mind and heart, whereas the singer who had only a good voice reached only the ears of the hearers. (Cooper in Hatton 1990: 53)

This "complicity," as Hatton refers to it, between listener and singer, concentrated the collective thoughts on one end. In a sense, the singer captured the audience by the power of his performance, and increased the collective power of the song (Hatton 1990: 54-55).

Margaret Laurent, in her work with Penobscot song and dance, claims that it was important that no word or syllable be spoken out of place.[11]

> So important was it that it be properly performed and the ritual song accompanying it be correctly rendered that any error in either dance or song was considered to invalidate the whole ceremony and it had to be performed over again. Otherwise, as they said, "the path would not be straight." (Laurent 1963: 8)

This capturing of others with song extended to other animals and conscious beings. In one legend, Kluskap charms a whale into carrying him across the water by singing. In another legend Ki'kwa'ju (wolverine), while being carried high into the sky by an eagle, sings a song to enchant the eagle to take him even higher (Rand 1971: 287; Wallis and Wallis 1955: 428).

Lonecloud described a song competition to "capture" the heart of a woman for a wife.

> The young Indian goes to the camp & sings songs. They were great singers. Another young Indian would come with his songs next night until the 7th one comes. On the 7th night she selects the one who will be her man. Then she is asked which of the singers she liked best. She says which she liked & that is her man. Some have love song, some warrior song, some bird songs, etc. & hunting songs etc. The Chief Medicine Man brings her choice & it is

pronounced then what time the marriage will be…. Warrior song—warrior dance, Great Spirit Song—Wild birds. Love song—what he will do when she gets in his canoe and paddles away. If she takes a liking to that then he's her man. When she selects her man she sings back and then its all over, and is pronounced by the Chief Medicine Man when the great marriage feast will take place. (Dennis 1923: Notebook 2: 77)

Alan Merriam, ethnomusicologist and cultural anthropologist, discusses how music embodies cultural beliefs, aesthetics and values, and is a form of socialization and education.

The confusion for most Westerners lies in the distinction between education and schooling; the lack of formal institutions in no way suggests that education, in its broadest sense, is absent…. Learning music is part of the socialization process…. (Merriam 1964: 146)

A variety of knowledge was conveyed in song texts. Songs were sung to communicate everyday information, they were sung for a particular purpose, and they were a specific form of social interaction.[12] Aside from *nskawaqn*, there were other types of song. *Ktapekiaqn* means "song," and in contemporary Mi'kmaw culture, it can refer to any song (e.g., a country and western song on the radio). *Ketu'muet* is the word for "someone who sings a chant," and it is usually done for someone who wants to dance or is dancing.

Wilfred Prosper said some of the songs were disparaging songs. "They have something against somebody. If you want to put on a certain wish to somebody." He also mentioned eloping songs (Johnson and Prosper, interview 1992).[13] The following is a divorce song translated by Francis.

It is her doing that we became married and now she mopes around.
It is her doing that we became married and now she cries, cries, cries.

This small list not only shows how information was embodied in song, it also indicates what type of information was important to sing. They were singing their lives. These songs, in turn, were shared with the community, and then became part of a body of knowledge passed down to subsequent generations.

The late Sarah Denny, a respected Mi'kmaw Elder and chanter who lived in Eskasoni First Nation, said that in her youth, women were not chanters, only men were. The women would sing and chant in their homes to teach their children. At that time (1920s and 1930s), she said, the priests told the people that it was wrong to sing and dance, and it began to die out. In an effort to preserve the chants and songs of her culture, Sarah obtained permission from the Grand Council to begin collecting, preserving and singing the chants and songs.

When you go to a Grand Council, they're chiefs, and you ask them to do what you want to do. Well, if you're good enough to chant for them and show them what you know, they'll really let you go ahead and do it and see what you could do with what you're doing—if you do your best anyway. So you get a permission to do it, then you're all ready to do it. (Denny, interview 1990)

Sarah Denny began visiting Elders to learn the chants, and later began to publicly teach and present the dances and songs with her children and others she trained. She described these chants as "gifts given by Elders" (ibid.).

CONCLUSION

The songs and chants of the Mi'kmaq discussed in this chapter are examples of mirroring the sounds and rhythms of the world, and of communicating information. Furthermore, they are a way of respecting these sounds and rhythms as messengers, and as acts of creation and destruction.

The sounds and rhythms of the world are reflections of the world itself. In the gull's song, children are taught about gulls sensing the change in atmosphere and responding with a specific call that the Mi'kmaq could use as an indicator for the weather. In the song of the owl, we see owls communicating with one another. This song then became a model for how people could communicate with one another. Even the stars have a song which scientists have begun to record. Some, as in the case of Johann Kepler, spent a lifetime documenting them. This type of mirroring the rhythms and patterns, just as a leaf reflects the rhythms and patterns in its structure, was being done by the Mi'kmaq through story, song and dance.[14] Not only did they remember, they passed these sounds on to the present day through songs and chants. It is another way sound travels.

Songs and chants were an effective mode of education used by the Mi'kmaq throughout time. Information was communicated through song and chant about the landscape. Songs were both formal and spontaneous, and sung for almost any occasion. Not only did they mirror the sounds and rhythms of the world, but they were another way to communicate information and education. They were/are creative, and were commonly given as gifts.

Song allows you to transcend boundaries imposed by ordinary speech. It can be refitted and shaped in innumerable forms, giving rise to endless forms of expression. It can be understood by anyone. Song reflected one's heart, it empowered, and it was used as a way to honour and give to another. It was also used to conquer. It allowed for a style of communication among people of a community, but it was a way for humans to listen to and learn about animals, birds and other living beings of the world. Song is a universal language.

> Songs are thoughts, sung out with the breath when people are moved by great forces and ordinary speech no longer suffices. Man is moved just like the ice floe sailing here and there in the current. His thoughts are driven by a flowing force when he feels joy, when he feels fear, when he feels sorrow. Thoughts can wash over him like a flood, making his breath come in gasps & his heartthrob. Something like an abatement in the weather will keep him thawed up. And then it will happen that we, who always think we are small, will feel still smaller. And we will fear to use words. But it will happen that the words we need will come of themselves. When the words we want to use shoot up of themselves—we get a new song. (Orpingalik 1995: 831-32)

CONCLUSION
THE STATE OF THE MI'KMAW LANGUAGE TODAY
(IT ISN'T GOOD)

> Even after thirty-eight years of the study of the Mi'kmaw language, I still feel enthusiastic enough to assist where I can with Mi'kmaw Language retention and promotion in any community from Newfoundland to Quebec, to the state of Maine, and throughout the Maritimes. I still feel there is so much to discover.
>
> Bernie Francis, 2011

This book is not the be-all and end-all. There is much yet to be researched and discovered. What has emerged from our research over the past twenty-plus years—and from the generosity of the Elders who were interviewed for this book—is the incredible richness of the Mi'kmaw culture, a culture that is still very much alive.

During this process, we have discovered how descriptive, flexible, precise and picturesque the Mi'kmaw language is. It adequately describes our world and the universe in a natural way that the Indo-European languages find much more difficult to express. It is our experience that the Mi'kmaw language has its own language of science, language of spirituality and language of governance and law, which were sung and danced into being. By the language of science we mean everything from astronomy, which observes and studies the heavenly bodies, to biology, which looks at smaller parts that make up the human animal body, plant structures, etc., and to traditional medicines, many of which are the basis of modern day pharmaceuticals.

However, if more drastic action is not taken by Mi'kmaw community members and educators, the Mi'kmaw culture and language are both in serious danger of being diminished to the point whereby they will no longer be recognized as a distinct Algonquian cultures. Though the language is extremely rich and expressive, it is becoming stale in the sense that it is not being updated. Unlike English, which dominates our world and experiences, and which has kept up with the changes, the Mi'kmaw language has not. The

Mi'kmaw language needs to be updated to reflect modern-day life because much has evolved as a result of the rapid societal changes.

Though there are some sincere attempts by a few Mi'kmaw communities to establish total Mi'kmaw language immersion programs, there are few programs in place to date. This slow progress to establish immersion programs is endangering the health and sustainability of the Mi'kmaw language; many eloquent and knowledgeable speakers of the language of this land are leaving this world, their knowledge lost. The Mi'kmaw language could easily go the way of Wampanoag, Passamoquoddy, Piquot and many other tribal languages. It is terrifying to think that Mi'kmaw is also heading in that direction despite our understanding of the importance of the retention of language, and despite strong research showing the effectiveness of immersion programs.[1] The thought and the possibility that this language may not exist fifty years from now is painful. We encourage all band councils to support all of their educators in promoting total immersion programs for each and every community where possible.

First language immersion programs are both successful and empowering to the cultures they represent and seek to preserve. The Maori of New Zealand began their immersion program more than thirty years ago, largely through the will and efforts of their own communities. The basis of their success is what they term Kohanga Reo ("language nest"). This is a system whereby they bring in infants to a school with parents and Elders who interact with these children using the Maori language. This approach has strengthened the Maori language to the point that they now have math books in Maori with margins using the English language to explain terms and concepts that either do not exist or have no equivalent in the Maori culture.

The Mi'kmaq are innovative enough to have a similar program that correctly reflects the Mi'kmaw way of life. There are still some excellent Mi'kmaw readers and writers who could act as resources but, above all, educators now have access to sophisticated technology that could assist us in total immersion programming. In addition, there are at least two dictionaries, one old, one contemporary, and a book that describes the grammatical structure of the Mi'kmaw language that teachers could easily access.[2]

Throughout this book, we have also given examples of how what are known today as the arts—song, dance, storytelling and visual arts were integral to how people learned, personalized this learning process and communicated their knowledge. Therefore, we would like to encourage different ways of learning through the arts, and in particular music which includes sounds, rhythms and movement.[3]

It appears to us that everything is in place. What we need is a strong will and conviction to move with great speed immersing Mi'kmaw children in the language of their ancestors, the language of this land, Mi'kma'ki. Perhaps this is the perfect time to begin to have serious community meetings with community members who are not necessarily involved in education. The basic questions put forth to the community members would be: "Do we wish our children to continue to be the keepers of the language? If so, are we willing to voice our total support to our chiefs and councillors to implement total immersion programs knowing that anything less would not be, and has not been, effective in maintaining the language?"

In the Mi'kmaw culture there are many *a'tukwaqnn*, best described as stories about Kluskap, largely regarded as a mythological character. But, the understandings of the Mi'kmaq are centred on the teachings left to us by the Kluskap stories, not as to whether he was real or not real. These teachings have their value in that they help the Mi'kmaw people reach a wise understanding of the world around them and teach us how to relate to it respectfully including other people who share the same surroundings.

Our understanding of these concepts—Kluskap and *wiklatmu'jk*—is really the language of our environment—wisdom, respect, spirituality and, in general, gratitude for what we have been allowed to use for our well-being on this planet; we see ourselves as part and parcel of the animal and plant life. The ongoing conditioning of Christo-Europeans to have "dominion" over the plants and animals, as written in the Christian Bible so long ago is antithetical to Mi'kmaw world view. After 300 years of this conditioning who wouldn't forget what their ancestors left them? But to act destructively, to destroy this land by way of pollution, overuse or overkilling, for example, is ultimately to destroy ourselves.

The Mi'kmaq do not wish to go the way of other cultures around the world, who may feel conflicted and confused by their spiritual identity because of differing interpretations of who and what they are. We reiterate these important factors this way: *wiklatmu'jk, Kluskap, kinap, jipijka'm*, etc., are all integral to our family—the family and culture of the Mi'kmaq. Important teachings of our world, the abundance of gifts that are contained therein and how we are expected to relate to it in a respectful and careful manner are embodied in what we term as energies, characterized by the Mi'kmaw language.

Weji-sqalia'timk—the Mi'kmaq sprouted or emerged from this landscape and nowhere else; their cultural memory resides here. People may argue that Kluskap and *wiklatmu'jk* did not exist; others may say they are real but irrelevant. Yet they embody teachings of how to relate with the world. Perhaps they are best looked upon as energies with varying degrees of power. They have and continue to serve a purpose to a people who have known and communicated with this land intimately since time immemorial. They are our family, they are part of our home and continue to be part of the cultural psyche of Mi'kma'ki.

AFTERWORD

It is my belief that many of the difficulties people are experiencing on modern day reserves are due to cultural amnesia—forgetting who we are and what a good community member and leader truly meant. Fortunately there are enough of our leaders out there who follow our treasured culture and who have not forgotten their core values, leaders sensitive enough to use their influence to re-educate their communities, especially the young who have aspirations of becoming leaders in the future.*

Yet money and power have become gods to many in leadership roles, while terms such as *weji-sqalia'ti'kw* and *ko'kmanaq*—which convey the sense of rootedness, sharing and being responsible for "all our relations"— seem to be fading into the past. At no time in our history have Mi'kmaw leaders been so much wealthier than their community members as they are today. The core values of Mi'kmaw culture, of sharing and being truly responsible for the well being of ALL community members, have become just fleeting traits exercised only when there is some potential personal gain for the leader. Sadly, many leaders are this way, even some who are worth millions, as has been attested to by media watchdogs as well as ordinary community members who are very aware of the facts but feel powerless to do anything.

As of late, I've been hearing the following phrase from our youth: "I want to be a leader someday so I can be rich." Sadly, this is clearly indicative of what they are seeing. What I would like to hear is "I want to be leader someday so I can help my people."

Mu na *wljaqo'ti* ansma wejianukw ta'n te'sunmn koqoey aq katu aunaqa ta'n *teli-tpi'j* kikmaq. Wljaqo'ti na me'j kijka' wjiatew ta›n koqoey pekwatu'n, katu *tetapu'-mimajuaqn* na wejiaq ta'n *teli-tpi'j aq teli-apoqnmaj* kikmaq.

As translated by Bernie Francis: *Happiness* is not so much in *having* as *sharing*. We make a living by what we get but we make a *life* by what we *give*.

Bernie Francis, 2012

About the Authors

Trudy Sable, PhD, is Director of the Office of Aboriginal and Northern Research at the Gorsebrook Research Institute, Saint Mary's University and an adjunct professor of Anthropology. She has been called a "muddy boots" community researcher and educator and has worked collaboratively for the past twenty-two years with First Nations and Inuit peoples within Canada and internationally. Her primary focus is on inter-generational, community/place-based, educational programs and research projects that bring Western science into dialogue with Indigenous Knowledge, promotion

of education through the arts and developing educational programs with the Innu Nation Environment Office. She is the SMU representative to University of the Arctic and to the Atlantic Association of Universities Working Committee on Aboriginal Issues. Dr. Sable has presented and published her research internationally. Along with Bernie Francis and representatives of the Mi'kmaw First Nations, she is currently working on the Pjila'si Mi'kma'ki: Mi'kmaw Place Names Digital Atlas and Website Project.

Bernie Francis, DLitt, grew up on the Maupeltu (Membertou) First Nation community in Cape Breton, NS. From 1970-1974, he worked as the Director of the Court Worker Program for the federal court system, a program that ensures fair and proper treatment for Mi'kmaw people including access to a translator. After leaving the court system, Bernie began his training in linguistics with linguist Doug Smith from the University of Toronto. Bernie completed that training in 1980, having developed a new orthography of the Mi'kmaw language with Professor Smith. The Smith/Francis orthography is now officially recognized by the Mi'kmaw chiefs in Nova Scotia, as well as by the Canada-Nova Scotia-Mi'kmaq Tripartite Forum. Dr. Francis received an honorary doctorate from Dalhousie University in October, 1999. He continues to work on many projects including the Pjila'si Mi'kma'ki: Mi'kmaw Place Names Digital Atlas and Website Project.

NOTES

Notes to Introduction

1. *Weji-sqalia'tiek* is a term used in a letter written on behalf of Mi'kmaw chiefs protesting the loss of their lands to the British, and submitted to the Governor of Halifax. It was summarized and translated by the Abbé Maillard and sent to the Supérieur des Missions Etrangères in Paris on October 18, 1749 (Maillard 1749: 17). Francis later translated the document and discovered the unique usage of this phrase. The phrase *saqaliaq* has been normally used only for leaves or plants and such, but never for humans.

2. "We exclusive" (*ninen*) in this case means "we, the Mi'kmaw people," but not others who came later from other countries or territories.

3. According to Bernie Francis, there have been, of late, some paper presentations and lectures at prestigious institutions on the concept of *netukulimk*. Having been raised with the culture, legends, language and spirituality, this concept has been overstated in its meaning.

> *Netukulimk* (infinitive form) is a concept referring to a way of gathering/ hunting for a living. This could mean anything from hunting large or small animals and birds, fishing, attending a blueberry and potato harvest in Maine, or money in the sense of bringing back a living. If people knew that you were leaving for the state of Maine for the blueberry harvest, they would normally ask you: "*Attukulin?*" meaning, "Are you going to return with 'abundance of wealth for living' (e.g., money). *Attukulimk* is also the infinitive form, but means "to actually go to make a living or bring support of some kind to benefit the family."

> Having brought the concept forward in the mid-1980s to the Government of NS at the time, its usage was essential to convince the department of lands and forests that the Mi'kmaw people understand only too well the meaning of conservation. In some of the recent presentations by scholars, *netukulimk* has been attributed with more meaning, including spiritual significance and such, than the actual meaning of the word.

4. For more information about the Debert site see Bernard et al. (2011).

5. "Nova Scotia has more than 400 lakes. The largest freshwater lake is Lake Rossignol, at the southern end of the peninsula. Larger than Rossignol is Bras d'Or, a saltwater tidal lake linked to the Atlantic Ocean and covering nearly one-fourth of Cape Breton Island. Rivers are numerous and short, generally less than 80 km (50 mi) long. All have their origins in the Atlantic Upland. The Annapolis, Sissiboo and Shubenacadie rivers flow into the Bay of Fundy. Among other rivers, the Mersey drains Lake Rossignol, and the Saint Mary's flows into the Atlantic Ocean east of Halifax." http://autocww.colorado.edu/~toldy3/E64ContentFiles/CanadianGeography/NovaScotia.html. Accessed 1 July 2011.

6. A number of discussions occurred during the research for this book, looking at the feasibility of Lewis's theory regarding the districts. Key people involved were William Jones, geomatics expert and geologist; Rob Ferguson, archaeologist for Parks Canada (now retired); Bernie Francis, Mi'kmaw linguist; Trudy Sable, anthropologist; David Christianson, archaeologist and Collections Manager, Nova Scotia Museum, as well as other staff at the Nova Scotia Museum.

7. There have been translations of Eskikewa'kik (district) such as "skin dressers," but this is not a satisfactory translation for linguist Bernie Francis at this time. We have noted it as "translation uncertain at this time."

8. W. H. Mechling has elaborated a whole network of kin terminologies that indicate a system of reciprocal relationships. Terms for brother and sister for instance, might extend to first cousins of the same sex on the side of the opposite-sex sibling of a parent (which would in turn include certain first cousins). This also would hold true of parental roles being extended to include certain nephews and nieces, who might be addressed as "son" or "daughter" (Mechling 1958: 75-101; McGee, lecture November 6, 1990).

Bernie Francis, however, feels that kin terminology as it is today, is a looser designation for people who one feels plays a particular role in one's life—like an uncle or father, or someone you might call grandmother because of her wisdom. A number of Mi'kmaq have mentioned this type of relationship they have with a person or persons in their community who are not literally their uncle or father, aunt, etc.

9. Anthropologist Harold McGee believes that early historical Mi'kmaw society followed a deme structure. In brief, he defines demes as "bilateral descent groups," that "approach corporate exclusiveness as kin groups" and "possess corporate exclusiveness as community groups" (McGee 1977: 109).

> Demes would seem to be ideal for organizing labour under conditions which normally only require the coordination of large task groups at certain specified times. The construction of fishing weirs and traps for the spring and autumn runs of anadromous fishes would be one such task, game drives would be another, and undoubtedly there are others. The majority of production activities would require fewer people and the resources would be more efficiently exploited with dispersed households scattered about the periphery of a bay, or along the banks of rivers. The compounds of prestigious headmen

would be larger than those of others in order to accommodate the bachelors (*ulbadoo* [*l'pa'tu* S/F]) who were normally attached to his household. The community would be the collection of scattered, bilaterally related households who recognized a common headman, and whose corporate identity was symbolized by a particular animal. (111-12)

Notes to Chapter 1

1. E.g., Wallace Bernard from Membertou, who had an amazing agility for playing on the double meanings.

2. In 1994, Doug Smith and Bernie Francis received an award from the Native Council of Nova Scotia in recognition for their work on the Mi'kmaw language. During this event, Trudy Sable organized a video session (September 30, 1994) with both Smith and Francis to discuss the language with a number of invited guests.

3. We know that by the 1750s, Abbé Maillard, a French Recollet priest among the Mi'kmaq was using the term Niskam. It is still used today in the prayers, such as the Lord's Prayer, that were translated into Mi'kmaw.

4. This translation is unedited and the Mi'kmaw terminology has not been transposed into the Smith/Francis orthography.

5. Badger is a mistaken translation of *ki'kwa'ju*, which means "wolverine."

6. *Mkamlamun* is an older form of the word indicating "one's heart." The "m" at the beginning of the word was used by the 19th-century missionary Pacifique, and existed at one time in the Mi'kmaw language. This "m" makes the word independent, or inalienably possessed. Nowadays, hardly anyone will use this form with "m" preceding it. If a person wants to refer to the heart, without indicating anyone's heart in particular, they would never begin with an "m." The speaker would more likely use *wkamlamun wen* to indicate "one's heart."

7. Trudy Sable found that while listening to the tapes of an interview later on, it became evident that the Elder had provided an incredible amount of information that embellished the point which was sought; he or she had provided context and shared his/her experience (see Sable 1996: 104).

Notes to Chapter 2

1. It has been suggested in recent years that a system of hieroglyphics among the Mi'kmaq predated contact, in which case it would be the first writing system indigenous to North America. We maintain that in fact these were mnemonic devices made by individuals, but they could not be read by everyone. They were not, as some would suggest, sacred writings. When Mi'kmaw travellers made marks on sticks and rocks, it meant something to that individual only.

2. Punk: soft crumbly wood that has been attacked by fungus, used as tinder.

3. As will be discussed in the next chapter, this mountain (Salt Mountain) is called Wi'sikk and is considered a "weather mountain" or a mountain the Mi'kmaq used

to predict the weather and had to be treated with respect. Parsons also included the following footnote from Rand in this account:

"There is also a reference to a distinctive personage Göminawe'nu, Grandmother of all, 'as Gluskap was our grandfather'" (Parsons 1925: 86 n. 7: Cp. Rand1: 293, 451).

4. The experience of a stone or rock as having some sort of consciousness, power or sense of the animate, combined in some cases with the physical attributes of the stone contributes to animacy being ascribed to it. Sometimes English language assigns gender to an object, such as the referring to ships as "she" or other personification, but this is more a tendency to project anthropomorphic qualities on objects than to truly experience them as conscious beings in their own right.

5. On April 21, 2010, Trudy Sable visited this site with Roger Lewis, Geomatics expert William Jones and photographer Joe Szostak. Due to Lewis's knowledge of the Shubenacadie River system, he was able to take us to this site and discuss why it would have been important to the Mi'kmaq.

6. Similar in theme, the following creation myth, recounted by Jerry Lonecloud to Clarissa Archibald Dennis in 1923, describes Kinap blowing life into the stone image, thus bringing it to life. The following is the unedited version of his story with a few corrections added in brackets for clarification:

> Ginup [Kinap] was greater than Glooscup [Kluskap]. Found stone image that was like a person. Come up to it and it looked so like a man he spoke to it and asked, "What are you doing here?" No response. Asked the second time and no response. Third time he stooped and blew his breath in images' mouth and he came to. He said, "Sit up," and he sat up. "Stand," and he stood up. "Walk," and he walked. "Stop," and he stopped. He named him Gloosap. Ginup says to Glooscup, "I have not finished this place. There is a lot of riley water." Wanted water clean for fish.... [The story continues on about Kluskap's role in creating the world.] (Dennis 1923: Notebook 1: 80)

7. There are many dialectical variations among Mi'kmaq in New Brunswick, Quebec and Nova Scotia. *Puklatmu'jk* (NB and QC) and *wiklatmu'jk* (NS) are examples of this variance.

Notes to Chapter 3

1. Elsie Clews Parsons told how Mrs. Morris took her away from the crowd and would not let anyone listen as she told Parsons her stories. This may have been because it was not the appropriate time or place. Parsons also noted a story, "Ugluchopt: Thunderbirds," told to her by a woman informant. After finishing the story, the woman asked her to not tell anyone else on the Island (Chapel Island) that she had told this story (Parsons 1925: 71). We may only speculate the reasons for her discreetness—perhaps it was frowned upon to share certain stories with "outsiders."

2. This story and subsequent analysis was originally used as an example for how a Mi'kmaw legend could be used to teach a science lesson using Mi'kmaw

traditional knowledge. The full discussion can be seen in Sable (1996). It is also the subject of an illustrated children's book entitled *Muin aqq L'uiknek te'sijik Ntuksuinu'k / Muin and the Seven Bird Hunters* (Marshall et al. 2010).

3. In the early 1990s, Sable conducted research into Mi'kmaw legends associated with lithic resource areas and workshop sites in Nova Scotia around the Minas Basin and into the Bay of Fundy. Part of the research was on the use of rocks and minerals in traditional Mi'kmaw culture for the development of a Native Science pilot program in cooperation with the Mi'kmaq Education Authority (now Mi'kmaw Kina'matnewey) and for the Department of Energy Mines and Resources. Having already conducted research into the dances of the Mi'kmaq, and researched Mi'kmaw traditional land use practices for Parks Canada, the importance of both oral and non-verbal modes of communication within the culture was evident. A natural step was to begin linking the legends to specific sites around the Atlantic Provinces and their association with important resource areas (Sable 1992; 1996).

4. According to Sable:

> Three sources were particularly important in this discovery. The first were Mi'kmaw friends and acquaintances who travelled through woods and waterways with me, or sat around kitchen tables talking about the landscape, their language and their culture. The second was the reading and hearing of various legends such as those recorded by Silas Rand, Elsie Clews Parsons and Stansbury Hagar. Most auspicious was the discovery of the journals of Clarissa Archibald Dennis, a Nova Scotian travel writer, in the Public Archives of Nova Scotia by a fellow researcher, Marilyn Harnish Moore. Marilyn informed me of the extensive interviews Dennis had conducted with a Mi'kmaw by the name of Jerry Lonecloud, a.k.a. Jeremiah Alexis Bartlett, who lived from 1854-1930. In these journals Lonecloud recounted numerous legends and information about the Nova Scotian landscape. The third, was the essential work of Mi'kmaw linguist, Bernie Francis. Francis translated and re-wrote the place names in the Smith/Francis orthography and introduced me to the fluidity and verb-like nature of the Mi'kmaw language. (Sable 1996, 1998)

A number of other archaeological, geological and ethnological resources were also helpful, but the telling moment that legends may encode maps came while reading Michael Deal's archaeological report on the lithic resource areas around the Minas Basin. The sites that Deal pinpointed on his maps were almost identical with the ones in the legends.

5. The Smith/Francis orthography is generally accepted throughout Nova Scotia, and officially adopted by the Tripartite Forum.

6. See, for instance, Parkhill (1997).

7. Though Rand translates this as "crossing over place," it literally means "spine."

8. "This was at Scotch Bay, where he chased the moose down. Spense's Island is the kettle upside down in which he cooked the moose. His dog is there in stone too. And this broken canoe is turned to stone" (Parsons 1925: 86). In footnote 5 to

the legend recorded by Parsons entitled "Gluskap and Beaver" she writes, "Digby Gut was dug by Beaver to escape. He had no time to go to the dam, went right through" (Lucy Pictou, Lequille, NS) (Parson 1925: 86).

9. Running parallel to the northern border of the Southern upland, and separated from it by the long, straight valley of Annapolis and Cornwallis Rivers, is a high range of lava or "trap" rock, which has been known for three centuries as North Mountain. Like a great palisade, enclosing the fertile valley, it reaches from Cape Blomidon southwestward along the edge of the bay of Fundy for 120 miles to Brier Island. At its northeast end it is separated from the Cobequid range by Minas channel and the drowned Minas basin.... The last structural feature of North mountain to be considered is the semicircular hook at its eastern end, which encircles Scotsman Bay [Scots Bay] separating Minas basin from the Bay of Fundy. This hook is the result of warping or "dishing" of the trap-sheet, which, after running without marked change of dip for 120 miles suddenly curves northward around the nose of the dish-like fold, whose axis lies beneath the waters of Minas channel. From "the lookoff" above Canning around the bend of North mountain at Cape Blomidon, the crest-line steadily falls, and the trap belt narrows until it tapers to a point at Cape Split. It emerges again only in a fragment at Cape d'Or across the channel. The triangular form of Minas basin and Cobequid Bay reflects closely the eastward extension of this dish or fold, the axis of which passes eastward from Cape Blomidon to Truro, where the soft red sandstone that underlies the trap is broadly exposed. That the structure is somewhat complicated is shown by additional outcrops of trap at Partridge Island, Five Islands [Moose Island is one of the Five Islands] and Bass River—probably outlying patches of the same trap-sheet brought down by minor folding and faulting. (Goldthwait 1924: 18, 22)

10. It is significant that Fundy shore chalcedony is rarely encountered east of the Shubenacadie River (cf. Nash 1986: 29,39), while chalcedony from the Cape d'Or/ Parrsboro area have been identified at the Paleoindian site at Debert (MacDonald 1968) and eastwards into Pictou County.

The Scots Bay sources are most easily accessed by water from the Minas Basin area, and the most abundant use of these chalcedonies is along the Minas Basin and up the Gaspereau River to the Gaspereau Lakes. The southward distribution of Fundy shore chalcedony seems to follow well known historic portage routes to the Atlantic, namely, via the Shubenacadie and Musquodoboit rivers in central Nova Scotia and via the Lequille and Mersey rivers in southwestern Nova Scotia (cf. Deal et al. 1987). The source areas on Digby Neck have not yet been surveyed by archaeologists. They may have provided the chalcedonies for inhabitants of the southwestern coast and may also have moved along the Lequille-Mersey route via Bear River.

The area of distribution of this premier quality lithic material within western Nova Scotia may be a reflection of the late prehistoric socio-political organization in the region. In particular, it corresponds closely to one of the seven Mi'kmaw political districts, the western district known as Kespoogwit, stretching from Cape Sable, at the southwestern tip of the province, to the Shubenacadie and Musquodoboit rivers (cf. Anderson 1919; Biard 1959: 89; Speck 1922: 93-105; Deal 1989: 3-5).

11. Lands for the Franklin Manor reserve in Cumberland County were set aside in 1865. In reports to the Department of Indian Affairs, Indian Agent A. T. Clarke referred to the area where Mi'kmaq were living as Parrsboro County (1880-81), then Halfway River (1881-82 report). A reference was also made to "Indians of Cumberland County who reside on the reserve of Franklin Manor on Hay Lake" (1884 Indian Agent report). Hay Lake was most likely Lake Newville, where Elder Doug Knockwood's family resided.

12. What brought this area to light occurred in October 2005 at the Debert Workshop, a multi-disciplinary conference about the ancient PaleoIndian site in Debert, Nova Scotia, hosted by the Confederacy of Mainland Mi'kmaq. The week before the conference, during a presentation of Sable's "Legends as Maps" research to an Elder's Advisory Board for Debert, discussion took place about this "Boar's Back." Elder, Doug Knockwood, a member of the advisory board, mentioned he grew up on Lake Newville, and that his grandfather owned land there during his lifetime. He also mentioned that there was a reserve named Franklin Manor, and that his uncle was one of the last people to live on the reserve.

13. Isle Haute, had not been extensively excavated during research in the mid-1990s but since then has yielded further supporting evidence of Mi'kmaw presence and ancient use of lithic resources. Isle Haute, as has already been seen, is cited in the legends concerning the Minas Basin. Rand recorded another legend told by Stephen Hood in 1869:

> In cutting open a beaver dam at Cape Chignecto, a small portion of the earth floated away; and Glooscap changed it into a moose and set his dogs on it. The moose took to the bay and made off; whereupon Glooscap turned him back into land, made him an island—the Isle of Holt—and fixed him there. He changed the dogs into rocks, which may be seen to this day, seated on their haunches, with their tongues lolling out of their mouths; the plain is called Ooteel (his dogs). Spenser's Island is his kettle turned over; and the scraps he shovelled out when trying out his oil still lie scattered around, but turned into stone. (Rand 1971: 236-37)

David Keenlyside's report from 1997 on the archaeology of Isle Haute notes John Erskine's first reporting of the site in 1955, specifically "a large archaeological site situated on the eastern tip of Ile Haute" (Keenlyside 1997: 1). As well, there is a location on the island named "Indian Flats." Copper fragments were also found.

An expedition led by David Christianson from the Nova Scotia Museum in the summer of 1997, revealed preliminary evidence that Isle Haute was used more as a lithic manufacturing site than a source of lithic materials (Christianson and Keenlyside 2000: 8). According to Christianson, the site can be dated at least to the Woodland period 600-800 years ago with a chance of late Archaic presence. One ridge-back ulu (a crescent-shaped cutting implement) dating to the Middle Archaic was dug up by a scallop trawler off the tip of Isle Haute (Christianson, personal communication 1997). Unfortunately, due to higher sea levels now inundating potential PaleoIndian and Archaic period sites, and natural and human disturbances to the area, earlier occupation may prove more difficult to determine, although earlier occupation sites have been dated at nearby Cape d'Or (Christianson and Keenlyside 2000: 9).

Christianson and Keenlyside also note that Isle Haute was most likely a "virtual cornucopia of marine life to harvest" and a favourable area to camp with fresh water and natural barriers as protection from the wind. According to Christianson, it may well have an important meeting half-way meeting place between New Brunswick and Nova Scotia in pre-contact times (*The Sunday Daily News* 1997: 5). There remain many unanswered questions about earlier occupation 8000-4000 years ago, but preliminary excavations suggest that the island was well known to the Mi'kmaq, an area used to manufacture lithic materials, as well as a multi-season harvesting area for marine life (Christianson and Keenlyside 2000: 8-9).

14. At Cape d'Or in the early 1980s, pre-contact occupations were dated to ca. 1200 and 2000 years ago (Christianson and Keenlyside 2000: 9). Keenlyside's work at Cape d'Or reveals that it was a work station for stone materials:

> ...there were lots of stone materials good for working that were close by and you find mound and mounds of material at this place where people have sat down. Rather than carry huge boulders away, they would work down material into convenient shapes and sizes and then take those that would be good for trading and working on later on. (Keenlyside 1997)

Cape d'Or is also known for its native copper, and was once part of the territory of the famed *sakimaw*, Membertou. Samuel de Champlain reported that Membertou gave King Henri IV Cape d'Or as a gift.

Ian Booth, an amateur geologist who has travelled extensively around the Minas Basin, also reported a large, high cave just past Indian Cove near Cape d'Or. From Rand's previously cited recording of the legend, "Wizard Carries Off Glooscap's Housekeeper" we see Cape d'Or is referred to as "Wigwam House." "He now pitched his tent near Cape d'Or, and remained there all winter; and that place still bears the name of 'Wigwam House'" (Rand 1971: 287). Nearby, Advocate Harbour was referred to as Kluskap's medicine Garden by Jerry Lonecloud so we begin to see the whole area as potentially offering resources for survival—shelter, food, hunting, lithics and medicine.

15. Trudy Sable has travelled to Kamestastin and other areas of Labrador within the Innu First Nation, and has been shown a few of these "weather" mountains.

16. A double-ended beaver incisor was excavated from the Augustine mounds in New Brunswick has been dated to 2300-2400 years old (Allen 1997).

17. McGee then illustrated this point by telling the Maliseet story of Babaloos recounted by Peter Perley of Tobique, New Brunswick:

> The story of Babaloos, the story about this plant that comes to the people to heal the people. The closing of that story is that after the people have been healed, they send off runners to the other communities to share the knowledge and as these runners were getting ready to leave, runners from

the other communities are coming in to share their knowledge. And it all happened the same night in all these communities. And I would agree with you [about stories being localized] because they all go down to a particular spot on the Saint John River where this plant grows to show the people in that locale where the plant grows. But the same story is associated with other locales on the Saint John River. So the story proper exists all over but it is adapted to specific locales. And part of the magic and mystery is that it took place simultaneously and this plant came to give this medicine, or to share the knowledge of this medicine simultaneously all over the place. So it doesn't become rooted in a particular place…. By allowing the story to change location, then it has survival significance. (McGee, personal communication 1996)

18. Another site mentioned by Allen is near Papineau Falls (Winpekijuik, "riley water") in northeastern New Brunswick where there is a rock with one long crack with a number of other cracks emanating from it. Gilbert Sewall recounted a legend associated with the rock passed down to him by his father. The legend tells how Kluskap caught a big salmon and threw the salmon on the bank leaving an impression of the back bone (Allen 1997; Sable 2011: 168).

19. A common metaphor for this sense of perception of the landscape, among some Aboriginal cultures, is the "Turtle Island" motif, a name for the continent of North America, thought of as the shell of a giant turtle surrounded by the oceans. This metaphor is not widely held among Mi'kmaq according to Bernie Francis.

Notes to Chapter 4

1. Anya Peterson Royce, in her book *The Anthropology of Dance,* states:

It is clear that dance utilizes a number of channels, the kinaesthetic, which is crucial to it alone of all the arts, and the visual, aural, tactile, and olfactory. Given the number and variety of channels, the potential of communication is quite strong. If all the channels are transmitting the same message, then the impact is multiplied by a factor of five. It is perhaps this capacity to assault all of one's senses simultaneously that makes dance such a potent, often threatening, vehicle of expression. (Royce 1977: 200)

2. In her essay "The Magic Circle," Suzanne Langer speaks about dance in terms of the notion of a "mythic consciousness" as distinct from "scientific consciousness." "Mythic consciousness" is the experience of reality that can not be scientifically measured or literally defined; symbol and reality are inseparable.

Paintings, sculpture, and literature … show us these Powers already fixed in visible and describable form…. The first recognition of them is through the feeling of personal power and will in the human body and their first representation is through a bodily activity which abstracts the sense of power from the practical experiences in which that sense is usually an obscure factor. This activity is known as "dancing." The dance creates an image of nameless and even bodiless Powers filling a complete, autonomous realm, a "world." [I]t is an envisagement of a world beyond the spot and the moment of one's animal existence, the first conception of life as a whole—continuous, superpersonal life, punctuated by birth and death, surrounded and fed by the rest of nature.... To the "mythic consciousness" these creations are realities, not symbols; they are not felt to be created by the dance at all, but to be invoked, adjured, challenged, or placated, as the case may be. The symbol of the world, the balletic realm of forces, is the world, and dancing is the human spirit's participation in it. (Langer 1983: 38-39)

3. Anya Peterson Royce, describes dance as "the human body making patterns in time and space." Royce goes on to state that to distinguish dance from other movement activities, a more inclusive definition of dance is necessary. Therefore, she elaborates to define dance as "rhythmic movement done for some purpose transcending utility" (Royce 1977: 1, 5).

Like language, dance can be discussed specifically in terms of its component parts—its individual movements, rhythm, spatial patterning, use of gravity, intensity and stasis—and how those parts are pieced together into a dance. Judith Lynne Hanna, in her book *To Dance is Human* writes:

> [Dance is] human behaviour, composed from the dancer's perspective, of 1) purposeful, 2) intentionally rhythmical, and 3) culturally patterned sequences of 4) non-verbal body movements other than ordinary motor activities, the motion having inherent aesthetic value. (Hanna 1988: 19)

Hanna then defines the notion of "aesthetic" as "cultural ideas of appropriateness and competence which guide evaluation, accomplished by rapt attention and contemplation" (Hanna 1988: 19). In a more detailed explanation, Hanna outlines dance as a specifically human activity based on biological characteristics, but what dance means within each culture is a different question. As has been discussed, in Mi'kmaw culture there was a psychic fluidity between animal and human realm so the boundary of what it means to be human is more permeable, as seen in the discussion on language.

4. Wilfred Prosper of Eskasoni remembered the "pine needle" dance in one form. As a child he made pine needle clusters dance by putting them upside down on top of a wash basin or box. Then, by tapping the surface of the basin or box, the clusters of needles would be made to "dance." (Johnson and Prosper, interview 1995). Among the Passamaquoddy, it is a women's dance.

5. Because dance occurs within a larger context, Royce speaks in terms of "dance event." For example, Royce points out that "in Mixtec, the word 'yaa' means dance, game and music" and that "the same adjectives are often applied to dance and music" (Royce 1977: 9).

6. Francis speculates that the actual word may been *ulaqn* meaning "dish" or "bowl," but may have been written as *oorakin* by early writers among the Mi'kmaq because birch bark was what they used to make these dishes.

7. This description alone could be treated as a separate book; it provides a vivid

example of the multisensory contextual nature of dance from the visual cues to the smells, tastes and the movements themselves.

The social order is also evident in the seating, the order in which each a person speaks and dances, and the entry of women, always with the eldest leading. Simultaneously, despite the order, everyone is given their turn to offer their dance and sing their praises and thanks to the host. Each person dances until they have exhausted themselves, displaying their strength. The dancers are encouraged and approved by those assembled. The honouring of the ancestors, the thanking of the host and the exchange between those assembled and the dancer(s) are all present (Sable 1996: 236-37).

8.   The use of repetition, hyperbole, the intensity of movement and the chants all heighten the awareness of the message and plant it more firmly in memory. The various poses and gestures act as mnemonic devices, storing the story, in a sense, in the body. The pauses, the approvals, the handshakes, the eye gaze, all are interactive communication between dancer and audience. Those assembled, in turn, give their approval to the dancer, and encourage him or them to continue. Neither the men or women continue their dance without this approval.

Furthermore, the rhythm of the dance, its pauses, its crescendos, can also be seen. The intensity of the movement ranges from complete stillness to frenzied, intense dancing. Each part of the body, including the use of eye gaze, is used to depict a message to those assembled and the ancestors. The women too, have their distinct dance, gestures and chants. They spin on their heels, raise one arm up, and one down, and exclaim a guttural "heh." (Sable 1996: 238)

9. Jerry Lonecloud calls *meteteskewey* the "most mighty of medicines" (qtd. in Dennis 1923: Notebook 1: 25).

10. The serpent dance may have nothing to do with the rattlesnake, as Hagar infers, and was not necessarily adopted from the southwest where rattlesnakes are found, as scholars have surmised. There are no rattlesnakes in the Maritimes. The timber rattlesnake (*Crotalus horridus*) can be found in Southern Ontario, and the Massasaugas rattlesnake (*Sustrurus catenatus*) lives in Southern and Western Ontario. The timber rattler and the copperhead (*Agkisttrodon contortrix*) do reach as far as southern Maine, but little research on the rattlesnake has been done in Maine. According to John Gilhen of the Nova Scotia Museum, people in early days mistook garter snakes (*Thomnophis sirtalis*) for rattlesnakes because of checkerboard pattern on their backs that resemble rattlesnakes. The garter snakes also display the aggressiveness of a rattler because they coil and snap at you in defence. However, the garter snake has no rattle and is voiceless (Gilhen, personal communication 1996).

The moulting of snakes takes place once a year, and is associated with their growth. This moulting can occur either in the autumn, prior to hibernation, or the first thing in spring, after they come out of hibernation. When snakes moult depends on the range of temperature the species function within. Temperature essentially governs the actions of reptiles, determining when they emerge from hibernation when they feed when they nest and when their eggs to hatch. The Maritime garter snake exuviates in the spring (Gilhen, personal communication 1996; Sable 1996: 246-47).

11. The power of the *jipijka'm* horn is illustrated in the story "The History of Usitebulajoo" (Wsitiplaju). In the story, a group of hunters come upon Wsitiplaju and his sister camping in their territory. Uncertain what to do with the strangers, they consult the village elders. The council of elders advises that the hunters visit the strangers in hopes of rooting the horn of a *jipitjka'm* into Wsitiplaju's hair. This accomplished, the horn will adhere to his head, and then grow and entwine around a tree, ultimately entrapping and disempowering Wsitiplaju. The hunters succeed in their mission, and Utsitebulajoo becomes imprisoned by the horn. The boy is only freed, after other failed attempts by his sister to saw through the horn, by the application of red ochre to the horn. The red ochre immediately dissolves the power of the horn (Rand 1971: 53-58).

The story, "A Man Became a Tcipitckaam" *(Jipijka'm)* told by John Newell of Pictou Landing, further illustrates the connection of the *jipijka'm* with medicine:

> Two brothers were hunting. They saw a trench. "What is that?" one of them asked. It had been made by a Tcipitckaam. One of the men lay down in it. He became larger and larger and stronger. The other could not get him out of the cavity. He followed it down into the water. He came back and narrated a big story: Tcipitckaam is a female. The man went to the bottom of the lake, and there found a wigwam. He went in. There he saw an old man, a woman and a girl. A boy came in. "This is my son," said the woman. "Your brother-in-law came in only a few minutes ago," the woman said to her son. "All right."
>
> The people from whom these two brothers came were Micmac. Among them was a medicine man. The medicine man said, "If he sleeps with her under the same blanket, we cannot bring him back. If he does not do so, we can." The medicine man went out, dug a trench, put water in it, and placed medicine upon this water. He climbed a tree and trimmed off the branches. Soon he saw two big dragons approaching. The dragons made a big noise. One came to the tree where the medicine man was, curled around and around it, and thrust up his head in the middle of the coil. The medicine man said to the returned brother, "All right. That is your brother." He was now a big Tcipitckaam, and the brother could not go near him. With a wooden knife the medicine man cut off the creature's head, and removed the entire body of the man. His wife was beside herself with joy. She jumped and danced, and shouted. The medicine man gave the man medicine which caused him to vomit. The brother said: "When I tried to converse with him, he made a noise like a Tcipitckaam—he could not speak properly." If he had stayed there another day, it would not have been possible for him to come back. This is true. We know it because the old Indians have handed it on to us. (Wallis and Wallis 1955: 345-46)

Most likely this is a description of a man who was poisoned by taking the "left" or wrong medicine. Speculating further, it may have to do with picking a female plant versus the male, since in this case the *jipijka'm* is female. In so doing, he becomes unconscious, and goes to the underworld, the dwelling place of the *jipijka'm*. His revival is brought about by the shaman administering an antidotal medicine. Through the proper procedures, the shaman is able to call the *jipijka'm* to the upper world (back to consciousness), and liberate the man's body from her clutches. Poisoning can cause comas, convulsions and delirium, and affect a person's respiratory and circulatory system. The man's inability to speak properly, may have been due to the effects of the poison constricting his muscles, causing him to speak like a *jipijka'm* (Dickey 1986: 95).

12. Other descriptions of medicine dances add to the understanding of the Mi'kmaq relationship to the gathering of medicine. Jerry Lonecloud gave two descriptions of dances having to do with medicine.

> Bad spirit gives you disease. Indians believed greatly in prevention. Indians make medicine and go through ceremony with the medicine to drive(?) out the evil spirit before they take it, so the evil spirit can't leave any sting upon the body. Sick or well the medicine is taken two times a year; spring and fall, everyone took it. Medicine was made in secret. They danced around it. Ceremony the evil spirit doesn't like is performed and he can't get in them. (Qtd. in Dennis 1923: Notebook 1: 129)

> When medicine is gathered in summer and winter, it is put aside. Dance takes place in winter. Medicine man gets up a dance to thank Kluskap for the privilege of having medicine put away for the season and ask for his cure with the medicine. Medicine man leads off the dance until he is exhausted, then he pronounces all medicine gathered good. Last sometimes seven hours. If he can dance a certain length of time, medicine is good. If not, bad. (97)

Notes to Chapter 5

1. According to Lonecloud, it was woman who brought the birds to earth from the stars.

2. Berendt further cites the work of George Leonard saying:

> At the root of all power and motion, there is music and rhythm, the play of patterned frequencies against the matrix of time. More than 2,500 years ago, the philosopher Pythagoras told his followers that a stone is frozen music, an intuition fully validated by modern science [$E=MC2$]; we now know that every particle in the physical universe takes its characteristics from the pitch and pattern of overtones of its particular frequencies, its singing. And the same thing is true of all radiation, all forces great and small, all information. Before we make music, music makes us.... The way music works is also the way the world of objects and events works.... The deep structure of music is the same deep structure of everything else. (Leonard in Berendt 1991: 89-90)

3. Bernie Francis witnessed this cordial welcoming in a few Marae while in New Zealand in 1999.

4. The song texts are language behaviour, different from music sound. The relationship of language to music is obviously different from ordinary speech. Song texts conveyed information, the sounds of the world; they were and are a reflection of language behaviour in terms of how words and syllables were shaped to fit musical structure (Merriam 1964: 87-88).

5. Language clearly affects music in that speech melody sets up certain patterns of sound which must be followed at least to some extent in music, if the music-text fusion is to be understood by the listener. Bright comments that:

> languages display regular patterns of high-pitched and low-pitched syllables, loud and soft syllables, long and short syllables, and different languages give different emphases to these factors. Since patterns involving these elements of pitch, dynamics, and duration are also among the basic elements of music, it is at least a reasonable hypothesis that there may be some cultures in which features of spoken languages have played a part in conditioning the musical patterns of song.... Music also influences language in that musical requirements demand alterations in the patterns of normal speech. Thus language behaviour in song is a special kind of verbalization, which sometimes requires special language in which it is couched. (Merriam 1964: 187-88))

6. Bernard Hoffman's comments on legends collected by Elsie Clews Parsons seem to support Hatton's view.

> The Micmac concept of reincarnation is preserved for us in an important collection of Micmac folktales collected by Elsie Clews Parsons. Two of these tales deal with a being named *Waisis Ketdu'muwaji Chi'num* [*Waisisk Ketu'muaji ji'nm*: "animals bring back man"; This phrase is more accurately translated as "the man chants to the animals."] living a long distance from the Micmac in the land of the Supernaturals (and in one of the tales at least seeming to represent Kluskap). The first account tells that some adventurers ... came to a wigwam with two doors. They wanted to stay overnight. "No, we can't keep you overnight," the man said, "We are busy overnight."—"What are you doing?" the man said.—"Overnight we are singing. The bones of the animals you have in the woods, I am singing for them to get their life back." He puts out the fire, he sings. He takes a moose bone. The moose jumps out. Caribou, mink, all come back to life. This man Waisis ketdu'muwaji chi'num (animals bring back man) makes them all alive again. (Parsons in Hoffman 1955: 372-73)

7. A number of comments by early chroniclers mention that Mi'kmaq had good voices but deride the sound of Mi'kmaw chanting "because they do not observe any regularity or measure, except such as their caprice may inspire (Le Clercq 1968: 292). "Innocent Indian racket" was another phrase used by Le Clercq to

describe the sound the women made during one dance (Le Clercq 1968 [1691]: 294). Most likely the atonal quality of the natural sounds of the world did not suit his ear.

Another description by Le Clercq also indicates a difference between the chanting styles of women and men, at least in some contexts. As with many of the accounts, the specific dance is not given, but perhaps it is a snake dance.

> Men force from their stomachs certain tones of ho, ho, ho, ha, ha, hé, hé, ho, ho, ha, he, he which pass for airs alike charming and melodious among our Gaspesians. (The women) ... do not force from the bottoms of their stomachs, as do the men, those hues and cries of ho, ho, of ha, ha, of hé, hé; but their only sound is made with their lips, and a certain hissing like sound. (Le Clercq 1968: 292-94)

8. From the scientific standpoint, "Experts have shown that the laws of bird music correspond to those of human music."

> Some birds—very much like humans—make "modern concert music"; in their music the "deviations" are frequent, as in the song of the blackbird, the "quintessential composer" among birds. The blackbird, writes [Rudolf] Haase, "sings highly complex melodies that are almost atonal. In this connection it is significant that certain notations of the blackbird's song from the nineteenth century display a degree of complexity that was reached in human music only much later, as, for example, in the opera Salome by Richard Strauss." (Berendt 1991: 88-89)

9. This demonstration was videotaped by Trudy Sable in 1993 at the Eskasoni powwow.

10. Such utterances as "Youh, hou, hou," or "e!," signalize closure. In Mi'kmaw, the word (sound?) *aweia* is glossed "sigh, final exclamation of a song" (Pacifique's Grammar n.d: 227). Dièreville, on the other hand, had a different description of the closure of a song. "A certain vocal note like this: Houen, houen, houen, if one can express it, marks the cadence, and they pause from time to time to give utterance to the terrifying yells with which the dances always end" (qtd. in Hoffman 1955: 689).

11. This monitoring of dance and song has already been illustrated in the discussion on *nskawaqn* and in Maillard's account of a feast in the previous section on dance in which approval was necessary for dances and chants to be repeated by both men and women. Within familiar, accepted forms, new information could be sung and communicated.

> After this the bridegroom thanked them [the assembled guests] promising as much as, and more than, his ancestors; then the assembly gave again the same cry. Then the bridegroom set about dancing; he chanted war songs which he composed on the spot and which exalted his courage and his worth, the number of animals he had killed, and everything that he aspired to do. (Denys 1968: 409)

12. According to Merriam, songs and chants were stable not static, able to incorporate change (i.e., the white man's houses seen in the Beothuk songs). The creation of song texts were "...subject to public acceptance and rejection, and therefore part of a broad learning process which contributes, in turn, to the process of stability and cultural change" (Merriam 1964: 162).

There were songs for trade, love, divorce, meeting someone, hunting, medicine, honour, war, choosing a chief, welcoming, divorce, death, celebration, naming a child, humour, feasting, lullabies and for almost any occasion. In 1851, W. E. Cormack recorded a list of songs sung by the Beothuk of Newfoundland. These were told to him by Shawnawdithit, the last known surviving Beothuk. We think of this list as her death song.

> The Beothics have a great many songs. Subjects—are of whiteman, Darkness, Deer, Birds, Boats, Of the other Indians, Bears, Boots, Hatchet, Shirt, Indian Gosset, Stealing man's boat, Shells, Pots, Whiteman's houses, Stages, Guns, fire stones, wood or sticks, Birch rind, Whiteman's jacket, Beads, Buttons, Dishes, men dead, Whiteman's head, Ponds, Marshes, Mountains, Water, Brooks, Ice, Snow, Seals, Fishes &c Salmon, Hats, Eggs, &c. In the song two or three wigwams sometimes join. (Qtd. in Howley 1980 [1915]: 230)

People were and are known to find their personal songs, particularly through visions or dreams. Nicholas Smith notes that among the Penobscots "during the change from boyhood to man a youth found his song which was usually given to him in a dream." (Smith, unpublished manuscript, n.d.: 6). In Mi'kmaw culture, a boy supposedly became a man with the killing of his first moose. This was celebrated with songs, feasts and dances. Perhaps this was also a time for the boy/man to find his new song of manhood. In current Mi'kmaw culture, traditionalists sometimes receive songs while on fasts. "The Mi'kmaw Honour Song," sung at almost every occasion, came to its composer through a vision.

13. Merriam also noted groups of songs that were for social commentary, particularly about behaviour that was deemed inappropriate. Song was a way to bring public notice to a problem or a matter, and was a form of social control, not dissimilar from gossip (Merriam 1964: 197).

Similarly, songs were a way for the community to monitor a person's emotional state, and for a person to give voice to their emotions. Le Clercq relates that a man or woman might tend toward severe melancholy or suicide in the event that they had been shamed, or lost a spouse, friend or relative. In one account, he describes the melancholy of Gaspesian women.

> Nothing, however, has been effective up to the present in checking the mania of our Gaspesian women, of whom a number would miserably end their lives, if, at the time when their melancholy and despair becomes known through the sad and gloomy songs which they sing, and which they make resound through the woods in a wholly delorous manner, some one did, not follow them everywhere in order to prevent and to anticipate the sad effects of their rage and their fury. It is however, surprising to see that this melancholy and despair become dissipated almost in a moment, and that these people, however

afflicted they seem, instantly check their tears, stop their sighs, and recover their usual tranquility, protesting to all those who accompany them, that they have no more bitterness in their hearts. (Le Clercq 1968 [1691]: 248-49)

In another account, Le Clercq notes the use of song and dance when Mi'kmaq were experiencing famine (186).

14. If all the notes of an octave are graphically displayed with their particular angles (the same octave operation ... Johannes Kepler used in his famous work "De Harmonice Mundi"), the result will be the shape of a primal leaf. Which simply means that the interval of the octave, and with it the very possibility of playing a perceiving music, bears within itself the shape of a leaf ... that a plant is capable of executing within its blossom a precise division into three parts as well as into five parts ... here the shape of the third, there in that of a fifth, which—just as in music, structure the shape of the leaf as intervals. (Kayser in Berendt 1991: 90)

Notes to Conclusion

1. Statistics Canada Health. *Knowledge of an Aboriginal Language and School Outcomes*, by Anne Guevrement and Dafna Kohen (2010), who support the importance of language immersion programs in Level 1 education and its effect on L2 and L3 performance.

2. E.g., Hewson and Francis (1990).

3. This topic is extensively researched in Sable's MA thesis "Another Look in the Mirror: Research into the Foundations for Developing an Alternative Science Curriculum for Mi'kmaw Children" (1996), in which an earlier version of this research exists, and her PhD thesis entitled "Emerging Identities: A Proposed Model for an Interactive Science Curriculum for First Nations Students" (2005). Both theses delve into the use of Mi'kmaw traditional ways of knowing and learning as well as offering a model for a cross-cultural pedagogy for teaching in schools today.

REFERENCES

Allen, Patricia. 1997. Presentation on Mi'kmaw Ceramics. Paper presented at the Mi'kmaw Heritage Garden Workshop, Eel River Bar First Nation, New Brunswick, 27 May.

Allen, Patricia, Michael Nicholas and Fidèle Thériault. 2004. Rocks Provincial Park: The 1994 Archaeological Survey and Historical Inventory. New Brunswick manuscripts in archaeology 33E. Fredericton, NB: New Brunswick Department of Culture and Sport Secretariat.

Andrerson, W. P., compiler. 1919. *Micmac place-names in the Maritime Provinces and Gaspe Peninsula, recorded by S. T. Rand*. Ottawa, Surveyor General's Office.

Andrews, Thomas D. 1990. *Yamoria's Arrows: Stories, Place-Names and the Land in Dene Oral Tradition*. Yellowknife, NWT: National Historic Parks and Sites, Northern Initiatives, Canadian Parks Service, Environment Canada.

Bakker, Peter. 1988. *Basque Pidgin Vocabulary in European-Algonquian Trade Contacts*. Papers of the Nineteenth Algonquian Conference, edited by W. Cowan, 7-15. Carleton University.

Basso, K. H. 2000. Stalking with Stories. In *Schooling the Symbolic Animal: Social and Cultural Dimensions of Education*, edited by Bradley A. U. Levinson et. al. Lanham, MD: Rowman and Littlefield.

Berendt, Joachim-Ernst. 1991 [1983]. *The World is Sound: Nada Brahma: Music and the Landscape of Consciousness*. Rochester, VT: Destiny Books.

Bernard, Tim, Leah Morine Rosenmeier and Sharon L. Farell, eds. 2011. *T'an Wetapeksi'k: Understanding From Where We Come. Proceedings from the 2005 Debert Research Workshop, Debert, Nova Scotia, Canada*. Truro, NS: Eastern Woodland Press.

Biard, Pierre. 1959 [1616]. Relation of New France, of Its Lands, nature of the Country, and of Its Inhabitants. In *Jesuit Relations and Allied Documents*. Ed., Trans. Ruben Gold Thwaites, vol. 3 and 4. New York: Pageant Book Company.

Biggar, H. P. 1924. *The Voyages of Jacques Cartier*. Publications of the Pubic Archives of Canada, No. 11, Ottawa: F. A. Acland.

Booth, Ian. N.d. Rockhounding in Nova Scotia. Self published leaflet.

Christianson, David and David Keenlyside. 2000. A First Look. In *The Nova Scotia Msueum Isle Haute Expedition, July 1997, Curatorial Report.* Number 90 edited by R. G. Grantham, pp. 7-10. Halifax, NS: Nova Scotia Museum.

Cooper, John M. 1957. The Gros Ventre of Montana. Part II: Religion and Ritual. In *Anthropological Series* No. 16, edited by Regina Flannery. Washington, DC: The Catholic University of America.

Deal, Michael. 1989. The Distribution and Prehistoric Exploitation of Scots Bay Chalcedonies. Paper prepared for the 22nd Annual Meeting of the Canadian Archaeological Association, Fredericton, New Brunswick, May 10-13, 1989.

Deal, M., J. Corkum D. Kemp, J. McClair, S. McIlquham, A. Murchison and B. Wells. 1987. *Archaeological investigations at the Low Terrace site (BaDg2), Indian Gardens, Queens County, Nova Scotia.* Nova Scotia Museum, Curatorial Report 63: 149-228. Halifax, NS.

Deloria, Vine Jr. 1973. *God is Red.* New York: Dell.

———. 1995. *Red Earth White Lies: Native Americans and the Myth of Scientific Fact.* New York: Scribner.

Dennis, Clarissa Archibald. 1923. Journals of Clarissa Archibald Dennis. Public Archives of Nova Scotia. MG. 1, Vol. 2867. Notebooks 1 & 2.

Denys, Nicholas. 1968 [1672]. *Descriptions & Natural History of the Coasts of North America (Acadia).* New York: Greenwood Press. (originally published as Champlain Society Publication II).

———. 1983. The Difference That There is Between the Ancient Customs of the Indians and Those of the Present. In *The Native Peoples of Atlantic Canada: A History of Indian-European Relations,* edited by H. F. McGee, 38-44. Ottawa: Carleton University Press.

Dickey, Norma H., ed. 1986. *Funk and Wagnall's New Encyclopedia,* vols. 20, 21.

Goldthwait, J. W. 1924. *Physiography of Nova Scotia.* Memoir 10 (122) Geological Series. Geological Survey of Canada, Department of Mines.

Guevrement, Anne and Dafna Kohen. 2010. Stats Canada Health—*Knowledge of an Aboriginal Language and School Outcomes* (2010).

Hagar, Stansbury. 1895. Micmac Customs and Traditions. *American Anthropologist* 8:31-42.

———. 1896. Micmac Magic and Medicine. *Journal of American folk-lore* 9(34).

———. 1900. The Celestial Bear. *The Journal of American Folklore* 13(49): 92-103.

Hallowell, A. Irving. 1976. Ojibwa Ontology, Behavior and World View. In *Contributions to   Anthropology: selected Papers of A. Irving Hallowell.* Chicago and London: University of Chicago Press

Hanna, Judith Lynne. 1988 [1979]. *To Dance is Human: A Theory of Nonverbal Communication.* Chicago: The University of Chicago Press.

Hatton, Orin T. 1990. *Power and Performance in Gros Ventre War Expedition Songs.* Canadian Ethnology Service, Mercury Series Paper 114. Canadian Museum of

Civilization.

Hewson, John and Bernard Francis. 1990. *The Micmac Grammar of Father Pacifique.* Memoir 7 Algonquian and Iroquoian Linguistics, Winnipeg, MB.

Hoffman. Bernard Gilbert. 1955. The Historical Ethnography of the Micmac of the Sixteenth and Seventeenth Centuries. PhD thesis. University of California, Department of Anthropology.

Howley, James P. 1980. *The Beothicks or Red Indians. The Aboriginal Inhabitants of Newfoundland.* Toronto: Coles Publishing Company.

Keenlyside, David L. 1997a. The Archaeology of Isle Haute: A First Look. Unpublished report on file at the Office of Aboriginal and Northern Research. Halifax, NS: Saint Mary's University.

———. 1997b. Presentation on Mi'kmaw Archaeology. Paper presented at the Mi'kmaw Heritage Garden Workshop, Eel River Bar First Nation, New Brunswick, 27 May 1997.

Knockwood, Isabelle. 1992. Out of the Depths. *The Experience of Mi'kmaw Children at the Indian Residential School at Shubenacadie, Nova Scotia.* Lockeport, NS: Roseway Publishing.

Langer, Suzanne K. 1983. The Magic Circle: In *What is Dance: Readings in Theory and Criticism.* Oxford: The Oxford University Press.

Laurent, Margaret E. 1963. Indian Music and Dance. In Miscellaneous Papers No. 2. New Hampshire Archaeological Society, 1 May 1963.

Le Clercq, Father Chrestien. 1968 [1691]. New Relation of Gaspesia With the Customs and Religion of the Gaspesian Indians. Ed., Trans. William F. Ganong. New York: Greenwood. First published in 1910 by The Champlain Society, Toronto.

Le Blanc, Barbara. 1995. Dance: A Collage of Definitions. Handout given for lecture at Saint Mary's University, Halifax, NS, Spring, 1995.

Lewis, Roger and Trudy Sable. Forthcoming. The Mi'kmaq: Mi'kmakik Teloltipnik L'nuk—How the People Live in Mi'kma'kik. In *Native Peoples: The Canadian Experience*, edited by Christopher Fletcher and Rod Wilson. Don Mills, ON: Oxford University Press.

Little Bear, Leroy. 1975. Dispute Settlement Among the Nacirema. *Journal of Contemporary Law.* Salt Lake City, UT: University of Utah.

———. 2001. *Aboriginal Paradigms: Implications for Relationships to Land and Treaty Making.* Expert Report for the plaintiffs in Buffalo v. the Queen.

Lomax, Alan. 1968. *Folk Song Style and Culture.* American Association for the Advancement of Science. Publication 88. Washington, DC.

MacDonald, G. F. 1968. *Debert: A Palaeo-Indian Site in Central Nova Scotia.* Anthropology Papers No. 16. Ottawa, National Museums of Canada.

Maillard, Abbé Antoine Simon Pierre. 1758. An Account of the Customs and Manners of the Micmakis and Maricheets Savage Nations, Now Dependent on the Government of Cape-Breton. From an original French manuscript letter, never published. London: Printed for S. Hooper and A. Morley at Gay's Head, near Beaufort Buildings in the Strand.

Marshall, Lillian, Murdena Marshall, Prune Harris and Cheryl Bartlett. 2010. *Muin aqq L'uiknek te'sijik Ntuksuinu'k / Muin and the Seven Bird Hunters*. Sydney, NS: Cape Breton Universoty Press.

McGee, Harold. 1977. The Case of Micmac Demes. In *Actes du Huitieme Congres Des Algonquinistes*, edited by William Cowand. Ottawa: Carleton University.

Mechling, William H. 1958. The Malecite Indians, with Notes on the Micmacs. *Anthropologica* 7:1-160.

Merriam, Alan P. 1964. The *Anthropology of Music*. Northwestern University Press.

Mott, R. J. 2011. Palaeoecology and Chronology of Nova Scotia Relating to the Debert Archaeological Site. In, *T'an Wetapeksi'k: Understanding From Where We Come. Proceedings from the 2005 Debert Research Workshop, Debert, Nova Scotia*, edited by Tim Bernard, Leah Morine Rosenmeier and Sharon Farell. Truro, NS: Eastern Woodland Press.

Nash, Ronald J. 1986. Micmac: Economics and Evolution. Curatorial Report No. 57. Halifax. Nova Scotia Museum.

Nova Scotia Dept. of Natural Resources. 2011. Virtual Field Trip. Stop 4. Shubenacadie Wildlife Park: Shoreline Deposits of Glacial Lake Shubenacadie. http://www.gov.ns.ca/natr/meb/field/stop4.asp#centrecontent. Accessed 13 February 2012.

Orpingalik. 1995. Personal Statement. *The Georgia Review* Winter: 829.

Pacifique, Fr. 1935. *Les Pays de Micmacs*. Monastère de Sainte Anne de Restigouche, Restigouche, Quebec.

Parker, Mike. 1995. *Where Moose and Trout Abound*. Halifax, NS: Nimbus.

Parkhill, Thomas C. 1997. *Weaving Ourselves into the Land: Charles Godfrey Leland, "Indians," and the Study of Native American Religions*. Albany, NY: State University of New York Press.

Parsons, Elsie Clews. 1925. Micmac Folklore. *Journal of American Folklore* 38:55-133.

———. 1926. Micmac Notes. St. Ann's Mission on Chapel Island, Bras D'or Lakes, Cape Breton Island. *Journal of American Folklore* 39:460-85.

Province of Nova Scotia. 1989 [1984]. Natural History of Nova Scotia. Volume One "Topics and Habitats." Prepared by Maritime Resource Management Service Inc. and Griffiths-Muecke Associates.

Radcliffe Brown, A. R. 1964. *Adaman Islanders*. Glencoe, NY: Free Press.

Rand, S. T. 1850. *Short Statement of Facts Relating to The History, Manners, Customs, Language, and Literature of the Micmac Tribe of Indians in Nova Scotia and P.E. Island*. Halifax, NS: James Bowes & Son.

Rand, Rev. Silas Tertius. 1875. *A First Reading Book in the Micmac Language, Comprising the Micmac Numerals and the Names of the Different Kinds of Beasts, Birds, Fishes, and Trees &c of the Maritime Provinces of Canada. Also, Some of the Indian Names and Places, And Many Familiar Words and Phrases, Translated Liberally into English*. Halifax, NS: Nova Scotia Printing Company.

———. 1888. *Dictionary of the Language of the Micmac Indians who Reside in Nova Scotia, New Brunswick, Prince Edward Island, Cape Breton and Newfoundland*.

———. 1971 [1894]. *Legends of the Micmacs.* New York and London: Longmans, Green. New York: Johnson Reprint Corporation.

Royce, Anya Peterson. 1977. *The Anthropology of Dance*. Bloomington and London: University of Indiana Press.

Sable, Trudy. 1992. *Traditional Use Study*. Prepared for Parks Canada, Atlantic Region Office.

———. 1996. Another Look in the Mirror: Foundations for Developing an Alternative Science Curriculum for Mi'kmaw Children. MA thesis, Saint Mary's University, Halifax, NS.

———. 1998. Multiple Layers of Meaning in the Mi'kmaq Serpent Dance. In Papers of the 28th Algonquian Conference, vol 28, edited by David H. Pentland, 329-40. University of Manitoba.

———. 2004. Negotiating Change, Maintaining Continuity: Science Education and Indigenous Knowledge in Eastern Canada. In *Investigating Local Knowledge: New Directions, New Approaches*, edited by Alan Bicker, Paul Sillitoe and Johan Pottier, 169-187. Hants, U.K.: Ashgate Publishing.

———. 2005. Emerging Identities: A Proposed Model For An Interactive Science Curriculum For First Nations Students. PhD Thesis, University of New Brunswick, Department of Education.

———. 2006. Preserving the Whole: Principles of Sustainability in Mi'kmaw Forms of Communication. In *Unlearning the Language of Conquest. Scholars Expose Anti-Indianism in America*, edited by Four Arrows (Don Trent Jacobs), 166-89. Austin, TX: University of Texas Press.

———. 2011. Legends as Maps. In *Tan Wetapeksi'k. Understanding From Where We Come*, edited by Tim Bernard, Leah Morine Rosenmeier and Sharon Farell, 157-77. Truro, NS: Eastern Woodland Press.

Smith, Theresa S. 1995. *The Island of the Anishnaabeg: Thunderers and Water Monsters in the Traditional Ojibwe Life-World*. Moscow, ID: University of Idaho Press.

Speck, Frank Gouldsmith. 1922. Beothuk and Micmac. In *Indian Notes and Monographs*, edited by F. W. Hodge. New York: Museum of the American Indian, Heye Museum.

———. 1915. Penobscot Tales. Some Micmac Tales from Cape Breton Island. Some Naskapi Myths from Little Whale River. *Journal of American Folk-Lore* 28:52-77.

———. 1940. *Penobscot Man: The Life and History of a Forest Tribe in Maine.* Philadelphia: University of Pennsylvania Press

Speck, Frank. 1976. *Penobscot Man: The Life and History of a Forest Tribe in Maine.* New York: Octagon Books.

———. 1985. *A Northern Algonquian Source Book*, edited by Edward S. Rogers. New York and London: Garland Publishing.

Spicer, Edward H. 1994. The Nations of a State. In *American Indian Persistence and Resurgence*, edited by Karl Kroeber, 27-49 Durham, NC: Duke University Press.

Stea, R. R. and R. J. Mott. 2005. Younger Dryas glacial advance in the southern Gulf of St. Lawrence, Canada: analogue for ice inception? *Boreas*, 34 (3): 345-62).

Wallis, Wilson D. and Ruth Sawtell Wallis. 1955. *The Micmac Indians of Eastern Canada.* Minneapolis: University of Minnesota Press.

Wernert, Susan J., ed. 1982. *North American Wildlife: An Illustrated Guide to 2,000 Plants and Animals.* Pleasantville, NY. Montreal: The Reader's Digest Association.

Whitehead, Ruth Holmes. 1982. *Micmac Quillwork.* Halifax, NS: Nova Scotia Museum.

———. 1988. *Stories from the Six Worlds: Micmac Legends.* Halifax: Nimbus Publishing Limited.

Wildlife Education. n.d. Bears. In *Zoobooks*. San Diego, CA: Wildlife Education.

Interviews

Personal interviews conducted by Trudy Sable

Vivian Basque, 2 August 1990. Halifax, Nova Scotia.

Annie Battiste, with Marie Battiste, 6 January 1992. Eskasoni, Nova Scotia.

Joel Denny, June 1993.

Sarah Denny, CKDU radio interview, *Rock Meets Bone*, Spring 1990.

Bernie Francis, 5 January 1992; 22 August 1992; 23 July 1994; 27 July 1995; 30 August 1995; 14 March 1996.

Joey Gould, 19 December 1990. Afton First Nation Community.

Dr. Margaret Johnson, 26 July 1995. Chapel Island Reserve, Nova Scotia.

Dr. Margaret Johnson and Wilfred Prosper, 9 and 11 December 1995. Eskasoni, Nova Scotia.

Joe Knockwood, Union of New Brunswick Indians, 16 December 1991.

Doug Smith and Bernie Francis: Videotape Discussion on the Mi'kmaw Language. 30 September 1994. Ramada Inn, Dartmouth, Nova Scotia.

Events

Eskasoni Powwow, June 1994. Eskasoni First Nation Community.

Personal Communication (with Trudy Sable)

Patricia Allen, Archaeologist, Province of New Brunswick (now retired). Phone interview Spring 2006.

David Christianson, Nova Scotia Museum. Personal conversation, April, 1997.

Michael Deal, Personal telephone conversation, June 1994.

Vaughen and Shirley Doucette, December 1995; January 1996. Eskasoni Reserve, Nova Scotia.

Gordon Fader, Geologist at the Bedford Institute of Oceanography, Phone interview 1996.

Bernie Francis: 10 October 1995; 4 December 1995; 29 December 1995; 1 January 1996; 10 January 1996; 19 January 1996; 21 January 1996; 26 February 1996.

John Gilhen, Assistant Curator, Nova Scotia Museum of Natural History, telephone interview 1996.

Bob Grantham, Geologist, Nova Scotia Museum of Natural History, Phone interview 1996.

John Hewson, Department of Linguistics, Memorial University of Newfoundland, St. John's, Newfoundland: 31 October 1995; 7 November 1995; 28 November 1995; 5 December 1995; 7 December 1995; 14 December 1995; 19 December 1995; 20 December 1995; 22 December 1995; 31 December 1995; 3 January 1995; 10 January 1996; 13 January 1996; 15 January 1996; 16 January 1996; 18 January 1996; 25 January 1996; 29 January 1996; 1 February 1996; 19 March 1996.

David Keenlyside, personal conversation, June 1994.

Doug Knockwood, personal conversation, October 2005.

Harold McGee, Professor of Anthropology (now retired), Saint Mary's University, October 1995 and February, 1996 Halifax, Nova Scotia.

Bob Ogilvie, Curator of Special Places, Nova Scotia Museum of Natural History, telephone interview 1994.

Ralph Stea, Geologist, Natural Resources, Canada, phone interview February 1996; personal conversations, 31 July 2009; 15 December 2011.

Ruth Holmes Whitehead, Assistant Curator (currently Curator Emeritus), Nova Scotia Museum, personal conversations. April 1993 and October 1995.

Lectures

Lecture on Mi'kmaw Oral Tradition. Gorsebrook Research Institute, Saint Mary's University, Halifax, November 11, 1994. Child Help Initiative

Harold McGee. Class lectures, Native People of the Atlantic Provinces (Ant. 322). Saint Mary's University, 6 October, 13 November 1990.

INDEX